urr
P. fine
be. nay be re
Book reader.
and, books ma

2

IRN

392

Fore

FUNDAMENTALS OF HYDRAULIC ENGINEERING SYSTEMS

Prentice-Hall Series in Environmental Sciences

Granville H. Sewell, Editor

FUNDAMENTALS OF HYDRAULIC ENGINEERING SYSTEMS

NED H. C. HWANG

Professor of Civil Engineering
University of Houston

Prentice-Hall, Inc., Englewood Cliffs, NJ 07632

Library of Congress Cataloging in Publication Data

Hwang, Ned H C
  Fundamentals of hydraulic engineering systems.

  Includes index.
  1. Hydraulics.   2. Hydraulic engineering.   I. Title.
TC160.H86   627   80-28055
ISBN 0-13-340000-X

Editorial/production supervision and interior design by Mary Carnis
Manufacturing Buyer: Anthony Caruso

Printed in the United States of America

10 9 8 7 6 5 4 3 2 1

Prentice-Hall International, Inc., *London*
Prentice-Hall of Australia Pty. Limited, *Sydney*
Prentice-Hall of Canada, Ltd., *Toronto*
Prentice-Hall of India Private Limited, *New Delhi*
Prentice-Hall of Japan, Inc., *Tokyo*
Prentice-Hall of Southeast Asia Pte. Ltd., *Singapore*
Whitehall Books Limited, *Wellington, New Zealand*

To My Parents

Who Taught Me That Fundamentals Are Beautiful

CONTENTS

PREFACE

This book is a fundamental treatment of engineering hydraulics in S.I. units (Systéme International d'Unités). It is primarily intended to serve engineering students and their teachers as a textbook. It is also intended to serve practicing engineers who need a convenient reference to basic principles and their applications in hydraulic engineering problems.

Teaching hydraulics in engineering schools has changed tremendously during the last two decades. Such change was prompted by two concurrent developments. First, many problems in hydraulics that seemed to be computationally insurmountable previously are now being solved by high-speed computers. Second, reduced credit hours in the basic engineering curriculum have resulted in a more comprehensive but condensed program at the bachelor level. Both of these developments emphasize the teaching of fundamentals.

Hydraulics is an area of fluid mechanics in which many empirical relationships are applied to achieve practical engineering solutions. Experience has shown that many engineering students, possibly well versed in basic fluid mechanics, were having trouble dealing with practical problems in hydraulics. For this reason, a good hydraulic textbook, presenting a systematic and comprehensive approach to practical problems, was needed.

This book contains ten chapters. They provide a broad coverage of most of the common topics in hydraulics systems. The first chapter discusses the fundamental properties of water as a fluid. The second chapter presents the concepts of water pressure and pressure force on surfaces in contact with water. In this chapter the basic differences between the S.I. system and the British units are discussed. Chapter 3 presents problems of water flow in pipes and pipe networks. The system concept is emphasized after an introduction to the basic elements. Chapter 4 discusses the basics of water pumps. Examples are provided for pump selections as an integrated part of the pipeline system.

Water flow in open channels is presented in Chapter 5. Detailed discussions of wave phenomena in supercritical flow channels and design of supercritical transitions are included. The hydraulics of wells and seepage problems are treated in Chapter 6. This chapter also contains a section on sea water intrusion due to excessive pumping in coastal regions. Several common hydraulic structures are presented in Chapter 7.

The proper use of scaled models is an essential part of hydraulic engineering. Chapter 8 discusses hydraulic models and the laws of similitude, with emphasis on their physical significances. Chapter 9 discusses measurement of water velocity, pressure, and discharge in pipes and in open channels. The elements of hydrology, the basic concepts and applications of hydrographs, frequency analysis, and flood routing in open channels are presented in Chapter 10.

This book is designed for a one-semester, three-hour a week course (16 lecture weeks) in the undergraduate curriculum. No prerequisite in basic fluid mechanics is needed. The book can also be used for the quarter-system academic program (12 weeks) by omitting a few special topics such as water pumps (Chapter 4), hydraulics of wells and seepage (Chapter 6), and hydraulic structures (Chapter 7). Examples and homework problems are provided for the students after every major section.

Computer programs are presented only where they are deemed necessary to emphasize a concept, such as the Hardy-Cross method for water distribution in pipe networks. In other areas in which computers are commonly used for heavy computations, such as open channel surface profiles, water hammer analysis, and flood hydrograph computations, digital programs are not included because I believe that simplified versions of these programs have little real significance and a complete treatment of these programs is beyond the scope of this book. Many of these programs (e.g., HEC-1, HEC-2*) are readily available to the public, and workshops are offered periodically by several universities and by the Hydrologic Center, U.S. Army Corps of Engineers.

Ned. H. C. Hwang
Houston, Texas

*HEC-1 Flood Hydrograph Package, The Hydrologic Center, Corps of Engineers, U.S. Army.
HEC-2 Water Surface Profiles Package, The Hydrologic Center, Corps of Engineers, U.S. Army.

ACKNOWLEDGEMENTS

The first attempt to arrange my lecture notes into a textbook was made in 1974. My colleagues at the University of Houston, particularly Professors Fred W. Rankin, Jr. and Jerry R. Rogers, have provided valuable suggestions and assistance which formed the foundation of this book. Dr. Rogers also carefully reviewed the entire text after its completion.

I am also deeply indepted to my students, Drs. Travis T. Stripling, John T. Cox, James C. Chang and Po-Ching Lu. All provided assistance during the various stages of preparation. Ahmed M. Sallam, who used a first draft of this book as lecture notes in teaching a course in hydraulics, provided many suggestions. I am particularly indepted to Carlos Hita, who not only carefully proofread every draft of the text and continuously offered valuable suggestions, but also provided, with comprehension and competence, most of the sample problems used in the text.

My dear friend, Dr. David R. Gross, an ever-inquisitive physiologist (as was Jean Louise Poiseuille, 1799–1869) with a great interest in hydraulic matters, reviewed the text and offered many irrefutable criticisms.

During the preparation of this book, I was ill for some time. The continuous encouragement, loyalty, and love of Maria, Leon, and Leroy kept me on track during the dark hours that have since passed. To them also, I dedicate this book.

Ned H. C. Hwang
Houston, Texas

INTRODUCTION

Hydraulic systems are designed to control, conserve, or transport water. The planning, construction, and operation of hydraulic systems involve the application of fundamental engineering principles in fluid mechanics, soil mechanics, structural analysis, engineering economics, and many other related fields.

Unlike most other branches of engineering, each hydraulic project encounters a unique set of physical conditions to which it must conform. There is no standard solution for which simple handbook answers can be assessed. Hydraulic engineering itself relies on fundamental knowledge that must be applied to meet the special conditions of each project.

The shape and dimensions of hydraulic systems may vary from a small flow meter a few centimeters in size to a levee several hundred kilometers long. Generally, however, hydraulic structures are relatively massive as compared with the productions of other engineering branches. For this reason, the design of large hydraulic systems is usually confined to the topography of the land. It is not always possible to select the most desirable location or material for a specific system. Commonly, a hydraulic system is designed to suit the local conditions including topographic, geological, ecological and social considerations, and the availability of native materials.

The vital importance of water to human life justifies the fact that hydraulic engineering is as old as civilization itself. There is much evidence that hydraulic systems of considerable magnitude existed several thousand years ago. For example, a large-scale drainage and irrigation system built in Egypt can be dated back to 3200 BC. Rather complex water supply systems, including several hundred kilometers of aqueducts, were constructed to bring water to ancient Rome. Dujonyen, a massive irrigation system in Siechuan, China, built nearly 2500 years ago, is still in effective use today. The abundant knowledge developed from these experiences has been found indispensable.

In addition to the analytical approach, modern hydraulic system design and operation depend heavily on many empirical formulas that produce excellent results in water works. No better replacement for these formulas can be expected in the near future. Unfortunately, most of these empirical formulas cannot be analyzed by modern methods in fluid mechanics. In general, they are dimensionally not homogenous. For this reason, the conversion of units from the British system to the S.I. units, and vice versa, is more than just a matter of convenience. Sometimes, the rigorous form (for example, the Parshall formulas for water flow measurements) must, for practical purposes, be maintained with its original units. In these cases, all quantities should be converted to the original units specified by the formula for computations.

For the most part, this book uses S.I. units. However, examples are provided when it is necessary to demonstrate the use of an empirical formula which is derived for the British units and the corresponding S.I. version is not commonly available. A detailed table of conversion is also provided in on page 333.

1

FUNDAMENTAL PROPERTIES OF WATER

The word hydraulic comes from the Greek word *hydraulikos*, meaning water. Hydraulic systems are designed to deal with water at rest and in motion. The *fundamentals in hydraulic engineering systems*, therefore, involve the application of engineering principles and methods to the planning, control, transportation, conservation, and utilization of water.

The physical properties of water are important to proper solution of various problems in hydraulic engineering systems. For example, the density, the surface tension, and the viscosity of water all vary, in one way or another, with water temperature. Density is a fundamental property that directly relates to the operation of a large reservoir. Change of density with temperature causes lake water to stratify in summer with warmer water on the surface. During fall, the surface water temperature drops rapidly and sinks toward the lake bottom. The warmer water near the bottom rises to the surface resulting in fall overturn of the lake. In winter, when water temperature falls below 4°C (highest water density), lake surface freezes while warm water remains at the bottom. The winter stratification is followed by spring overturn of the lake. Similarly, variation of surface tension directly affects the evaporation loss from a large water body in storage; the variation of water viscosity with temperature is important to all problems involving water in motion.

This chapter discusses the fundamental physical properties of water that are important to problems in hydraulic engineering systems.

1

1.1 THE THREE PHASES OF WATER

The water molecule is a stable chemical bond of oxygen and hydrogen atoms. The amount of energy holding the molecules together depends on the temperature and pressure present. Depending on its energy content, water may appear in either solid, liquid, or gaseous forms. Snow and ice are the solid forms of water; moisture, water vapor in air, is water in its gaseous form; and the liquid is its most commonly recognized form. The three different forms of water are called its three phases.

To change water from one phase to another phase, energy must either be added or taken away from the water. The amount of energy required to change water from one phase to another is known as *latent energy*. This amount of energy may be in the form of heat or pressure. Heat energy is usually expressed in the unit of calories (cal.). One cal is the energy required to increase the temperature of 1 g of water, in liquid phase, 1°C. The amount of energy required to raise the temperature of a substance by 1°C is known as the *specific heat* of that substance. The latent heat and specific heat of all three phases of water are discussed below.

Under standard atmospheric pressure, the specific heat of water and ice are, respectively, 1.0 and 0.465 cal/g/°C. For water vapor, the specific heat under constant pressure is 0.432 cal/g/°C, and at constant volume it is 0.322 cal/g/°C. These values may very slightly, depending on the purity of the water. The melting of 1 g of ice, changing water from its solid to liquid phase, requires the latent heat of 79.71 cal. To freeze water, the same amount of heat energy must be taken out of each gram of water. The process is reversed. *Evaporation*, the changing of liquid-phase water into its gaseous phase, requires a latent heat of 597 cal/g.

Evaporation is a rather complex process. Under standard atmospheric pressure,* water boils at the temperature of 100°C. At higher elevations, where the atmospheric pressure is less, water boils at temperatures lower than 100°C. This phenomenon may be explained best from a molecular exchange viewpoint.

At the gas-liquid interface there is a continual interchange of molecules leaving the liquid to the gas and molecules entering the liquid from the gas atmosphere. Net evaporation occurs when more molecules are leaving than are entering the liquid; net *condensation* occurs when more molecules are entering than are leaving the liquid.

The continuous impingement of vapor molecules on the liquid surface creates a pressure on the liquid surface known as the *vapor pressure*. This partial pressure, combined with the partial pressures created by other gases in the atmosphere, makes up the total atmospheric pressure.

---

*Standard atmospheric pressure is 101,376 N/m². It is approximately equivalent to 10 m of water column.

As the temperature of the liquid increases, the molecular energy is raised, causing a large number of molecules to leave the liquid, which, in turn, increases the vapor pressure. Finally, the temperature reaches a point where the vapor pressure is equal to the ambient atmospheric pressure; evaporation increases drastically: boiling in liquid takes place. The boiling temperature of a liquid is commonly known as the *boiling point* of the liquid. For water at sea level, the boiling point is 100°C. The vapor pressure of water is shown in Table 1.1.

TABLE 1.1 Vapor Pressure of Water

Temperature °C	Vapor Pressure		Temperature °C	Vapor Pressure	
	atm	N/m²		atm	N/m²
−5	0.004162	421	55	0.15531	15745
0	0.006027	611	60	0.19656	19924
5	0.008600	873	65	0.24679	25015
10	0.012102	1266	70	0.30752	31166
15	0.016804	1707	75	0.38043	38563
20	0.023042	2335	80	0.46740	47372
25	0.031222	3169	85	0.57047	57820
30	0.041831	4238	90	0.69192	70132
35	0.055446	5621	95	0.83421	84552
40	0.072747	7377	100	1.00000	101357
45	0.094526	9584	105	1.1922	120839
50	0.12170	12331	110	1.4139	143314

In a closed system, for example, in pipelines or pumps, water vaporizes rapidly in regions where the pressure drops below the vapor pressure. This phenomenon is known as *cavitation*. The vapor bubbles formed in cavitation usually collapse in a violent manner, which may cause considerable damage to the system. Cavitation in a closed hydraulic system should be avoided by maintaining the pressure everywhere above the vapor pressure in the system.

PROBLEMS

1.1.1. How much heat energy must be removed from 500 l of water at 80°C in order to freeze it?

1.1.2. A shallow pan contains 1200 g of water at 45°C. Compute the heat energy (in calories) required to evaporate this body of water under ambient pressure of 0.9 bar.

1.1.3. A simple barometer, Figure 1.1, uses water at 40°C as the liquid indicator. If the liquid column rises to a height of $h = 8.7$ m in the vertical tube, compute the atmospheric pressure. What is the percentage error by direct reading?

8.7 m

Figure 1.1 Simple barometer.

1.1.4. A pump is installed on the ground level to remove water from a deep pit at 60°C. Determine the maximum distance the free surface in the pit is allowed to drop below the suction end of the pump such that the pump remains in effective operation. Neglect energy losses.

1.2 MASS, DENSITY, AND SPECIFIC WEIGHT

The *density* of a substance is defined as the mass per unit volume. It is a property inherent to the molecular structure of the substance. This means that density not only depends on the size and weight of the molecules but also on the mechanisms by which the molecules are bonded together. The latter usually varies as a function of temperature and pressure. Because of its peculiar molecular structure, water is one of the few substances that expands when it freezes. The expansion of freezing water, when contained in a closed container, causes stresses on the container walls. These stresses are responsible for the bursting of frozen water pipes, chuck holes in pavement, and for the weathering of rocks in nature.

Water reaches a maximum density at 4°C. It becomes less dense when further chilled or heated. The density of water is shown as a function of temperature in Table 1.2. Note that the density of ice is different from that of

TABLE 1.2 Density of Water

Temperature (°C)		0° (Ice)	0° (Water)	4°	10°	20°	30°	40°
Density	(g/cm³)	0.917	0.999	1.000	0.999	0.998	0.996	0.992
	(kg/m³)	917	999	1000	999	998	996	992

Temperature (°C)		50°	60°	70°	80°	90°	100°
Density	(g/cm³)	0.998	0.983	0.978	0.972	0.965	0.958
	(kg/m³)	988	983	978	972	965	958

liquid water at the same temperature. We observe this phenomenon when we see ice float on water.

Sea water contains salt which increases the density of the sea water to about 4% more than that of fresh water.

In the S.I. system the weight of an object is defined by the product of its mass (m, in grams, kilograms, etc.), and the gravitational acceleration ($g = 9.81$ m/s² on earth). The relation may be written as

$$W = m \cdot g \tag{1.1}*$$

Weight in the S.I. system is usually expressed in the force units of *newton* (N). One newton is defined as the force required to accelerate 1 kg of mass at a rate of 1 m/s². The specific weight (weight per unit volume) of water, γ, is thus the product of the density, ρ, and the gravitational acceleration, g. For water at 4°C, the unit weight is $\gamma = 9810$ N/m³ ($= 981$ dyn/cm³). The ratio of the specific weight of any liquid to that of water at 4°C is called the *specific gravity* of that liquid.

Example 1.1

A cylindrical water tank (Figure 1.2) is suspended vertically on the sides. The tank has a 2-m diameter and is filled with 40°C water to 1 m height. Determine the force exerted on the tank bottom.

Figure 1.2

Solution

The force exerted on the tank bottom is equal to the weight of the water body.

---

*In the British (Imperial) system the mass of an object is defined by its weight (ounce, pound, etc.) and the gravitational acceleration ($g = 32.2$ ft/s² on earth). The relation may be written as

$$m = W/g \tag{1.1a}$$

Mass in the British system is expressed in the units of *slug*. One slug is defined as the mass of an object which requires 1 lb of force to reach an acceleration of 1 ft/s².

$$F = W = m \cdot g = [\rho \cdot (\text{Volume})] \cdot (g)$$

$$= \left[992 \text{ kg/m}^3 \cdot \left(\frac{\pi \cdot 2^2}{4} \times 1 \text{ m}^3 \right) \right] (9.81 \text{ m/s}^2)$$

$$= 30,572.47 \text{ kg-m/s}^2 = 30,572.47 \text{ N}$$

PROBLEMS

1.2.1. A container has a 5-m³ volume capacity and weighs 1500 N when empty and 47,000 N when filled with a liquid. What is the mass density of the liquid?

1.2.2. The unit of energy in the British system is foot-pound. Convert one unit of energy in this system to the S.I. system.

1.2.3. The unit of mass in the S.I. system is gram. Convert one unit of mass in this system to the British system in slugs.

1.3 VISCOSITY OF WATER

Water responds to shear stress by continuously yielding an angular deformation in the direction of the shear. The rate of angular deformation is proportional to the shear stress, as shown in Figure 1.3. The schematic diagram shown in Figure 1.3 represents the physical basis of viscosity. Consider that water fills the space between two parallel plates at a distance y apart. A horizontal force T is applied to the upper plate and moves it to the right at velocity V, while the lower plate remains stationary. The shear force T is applied to overcome the water resistance R, and it must be equal to R since there is no acceleration involved in the process. The resistance per unit area of the upper plate (shear stress, $\tau = R/A = T/A$) has been found to be proportional to the rate of angular deformation in the fluid, $d\theta/dt$. The relationship may be expressed as

$$\tau \propto \frac{d\theta}{dt} = \frac{\frac{dx}{dy}}{dt} = \frac{\frac{dx}{dt}}{dy} = \frac{dv}{dy}$$

where $v = \dfrac{dx}{dt}$ is the velocity of the fluid element, or

$$\tau = \mu \frac{dv}{dy} \tag{1.2}$$

The proportionality constant, μ, is called the *absolute viscosity* of the fluid.

The absolute viscosity has the dimension of force per unit area (stress) times the time interval considered. It is usually measured in the unit of poise (named after the French engineer-physiologist J. L. M. Poiseuille). The

Figure 1.3. Shear stress in fluid.

absolute viscosity of water at room temperature (20.2°C) is equal to 1 centipoise, which is one hundredth of a poise.

$$1 \text{ poise} = 0.1 \text{ N} \cdot \text{s/m}^2 = 100 \text{ centipoise}$$

The absolute viscosity of air is approximately 0.17 centipoise.

In engineering practice it is often convenient to introduce the term kinematic viscosity, v, which is obtained by dividing the absolute viscosity by the mass density of the fluid at the same temperature: $v = \mu/\rho$. The kinematic viscosity carries the unit of cm²/s (with the unit of Stoke, named after the British mathematician G. G. Stoke). The absolute viscosities and the kinematic viscosities of pure water and air are shown as functions of temperature in Table 1.3.

TABLE 1.3 Viscosities of Water and Air

Temperature °C	Water		Air	
	Viscosity N·s/m²	*Kinematic Viscosity* m²/s	*Viscosity* N·s/m²	*Kinematic Viscosity* m²/s
0	1.781×10^{-3}	1.785×10^{-6}	1.717×10^{-5}	1.329×10^{-5}
5	1.518×10^{-3}	1.519×10^{-6}	1.741×10^{-5}	1.371×10^{-5}
10	1.307×10^{-3}	1.306×10^{-6}	1.767×10^{-5}	1.417×10^{-5}
15	1.139×10^{-3}	1.139×10^{-6}	1.793×10^{-5}	1.463×10^{-5}
20	1.002×10^{-3}	1.003×10^{-6}	1.817×10^{-5}	1.509×10^{-5}
25	0.890×10^{-3}	0.893×10^{-6}	1.840×10^{-5}	1.555×10^{-5}
30	0.798×10^{-3}	0.800×10^{-6}	1.864×10^{-5}	1.601×10^{-5}
40	0.653×10^{-3}	0.658×10^{-6}	1.910×10^{-5}	1.695×10^{-5}
50	0.547×10^{-3}	0.553×10^{-6}	1.954×10^{-5}	1.794×10^{-5}
60	0.466×10^{-3}	0.474×10^{-6}	2.001×10^{-5}	1.886×10^{-5}
70	0.404×10^{-3}	0.413×10^{-6}	2.044×10^{-5}	1.986×10^{-5}
80	0.354×10^{-3}	0.364×10^{-6}	2.088×10^{-5}	2.087×10^{-5}
90	0.315×10^{-3}	0.326×10^{-6}	2.131×10^{-5}	2.193×10^{-5}
100	0.282×10^{-3}	0.294×10^{-6}	2.174×10^{-5}	2.302×10^{-5}

PROBLEMS

1.3.1. A flat plate of 100 cm² area is being pulled over a fixed flat surface at a constant velocity of 75 cm/s. An oil film of unknown viscosity separates the plate and the fixed surface by a distance of 0.1 cm. If the force required to pull the plate is measured to be 51.7 N and if the viscosity of the fluid is constant, determine the viscosity.

1.3.2. A flat circular disk of radius 1.0 m is rotated at an angular velocity of 0.65 rad/s over a fixed flat surface. An oil film separates the disk and the surface. If the viscosity of the oil is 16 times that of water at 20°C and if the space between the disk and the fixed surface is 0.5 mm, determine the torque required to rotate the disk.

1.3.3. A rotating-cylinder viscometer consists of two concentric cylinders with a uniform gap between them. The gap of a known width is filled with the liquid to be measured. If the inner cylinder rotates at 2000 rpm and the outer cylinder remains stationary and measures a torque of 2×10^5 dyn-cm, determine the absolute viscosity of the liquid. The inner cylinder diameter is 5.0 cm, the gap width is 0.02 cm, and the liquid fills up to a height of 4.0 cm in the cylindrical gap.

1.3.4. Compare the ratios of the absolute and the kinematic viscosities of air and water at (a) 20°C and (b) 80°C. Discuss the differences.

1.3.5. In the British system the kinematic viscosity is measured in units of feet² per second. Convert one unit in this system to the S.I. units in stokes.

1.4 SURFACE TENSION AND CAPILLARITY

Even at a small distance below the surface of a liquid body, liquid molecules are attracted to each other by equal forces in all directions. The molecules on the surface, however, are subjected to an inward attraction which is not balanced by the outward attraction. This causes the liquid surface to seek a minimum possible area by exerting surface tension tangent to the surface over the entire surface area. A steel needle floating on the water surface, the spherical shape of dewdrops, and the rise or fall of liquid surface in capillary tubes are the results of surface tension.

Most liquids adhere to solid surfaces. The adhesive force varies depending on the nature of the liquid and of the solid surface. If the adhesive force between the liquid and the solid surface is greater than the cohesion in the liquid molecules, the liquid tends to spread over and wet the surface, as shown in Figure 1.4(a). If the cohesion is greater, a small drop forms, as shown in Figure 1.4(b). Water wets the surface of glass but mercury does not. If we place a smallbore vertical glass tube into the free surface of water, the water surface in the tube rises. The same experiment performed with mercury will show that the mercury falls. These two typical cases are schematically presented in Figure 1.5(a) and 1.5(b). The phenomenon is commonly known

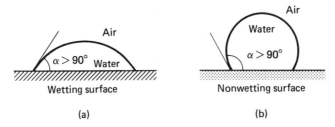

Figure 1.4. Wetting and nonwetting surfaces.

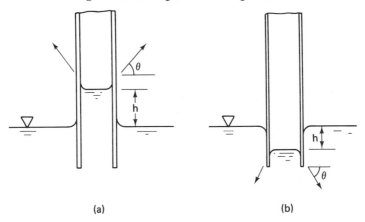

Figure 1.5. Capillary actions.

as *capillary action*. The magnitude of the capillary rise (or depression), h, is determined by the balance of adhesive force between the liquid and solid surface and the weight of the liquid column above (or below) the liquid free surface.

The angle θ at which the liquid film meets the glass depends on the nature of the liquid and the solid surface. The upward (or downward) motion in the tube will cease when the vertical component of the surface tension force around the edge of the film equals the weight of the raised (or lowered) liquid column. When the very small volume of liquid above (or below) the base of the curved meniscus is neglected, the relationship may be expressed as

$$(\sigma \pi D) \sin \theta = \frac{\pi D^2}{4}(\gamma h)$$

Hence,

$$h = \frac{4\sigma \sin \theta}{\gamma D} \tag{1.3}$$

where σ and γ are the surface tension and the unit weight of the liquid, respectively, and D is the inside diameter of the vertical tube.

The surface tension of a liquid is usually expressed in the units of force per unit length. Its value depends on the temperature and the electrolytic content of the liquid. Small amounts of salt dissolved in water tend to increase the electrolytic content and, hence, the surface tension. Organic matter (such as soap) decreases the surface tension in water, permitting the formation of bubbles. The surface tension of pure water is listed in Table 1.4.

TABLE 1.4 Surface Tension of Water (dyn/cm)

Temperature (°C)	0°	10°	20°	30°	40°	50°	60°	70°	80°	90°
σ		74.16	72.79	71.32	69.75	68.18	67.86	66.11	64.36	62.60	60.71

PROBLEMS

1.4.1. A small amount of solvent is added to the ground water to change its electrolytic content. As a result, the contacting angle, θ, representing the adhesion between water and soil material, is increased from 30° to 42° (Figure 1.6). If the soil has a uniform pore size of 0.7 mm, determine the change of capillary rise in the soil.

30° ~ 42°

0.7 mm

Figure 1.6

1.4.2. A liquid at 20°C is observed to rise in a 0.5-mm glass tube to a height of 18 mm. The contacting angle is measured to be 42.8° Determine the surface tension of the liquid if its density is 998 kg/m³.

1.5 ELASTICITY OF WATER

Under ordinary conditions, water is commonly assumed to be incompressible. In reality, it is about 100 times as compressible as steel. It is necessary to consider the compressibility of water when water hammer problems (Chapter 3) are involved. The *compressibility of water* is inversely proportional to its *volume modulus of elasticity*, E_b, also known as the *bulk modulus of elasticity*. The relationship may be expressed as

$$\Delta P = -E_b\left(\frac{\Delta Vol}{Vol}\right) \tag{1.4}$$

where E_b is the bulk modulus and Vol is the initial volume. ΔP and ΔVol are the corresponding changes in pressure and volume, respectively. The bulk modulus of water, as a function of both temperature and pressure, is shown in Table 1.5.

TABLE 1.5 Bulk Modulus of Water $E_b/10^9$ (N/m²)

Pressure × 10⁵ N/m²	Temperature			
	0°C	10°C	20°C	50°C
1 ~ 25	1.93	2.03	2.07	
25 ~ 50	1.96	2.06	2.13	
50 ~ 75	1.99	2.14	2.23	
75 ~ 100	2.02	2.16	2.24	
100 ~ 500	2.13	2.27	2.34	2.43
500 ~ 1000	2.43	2.57	2.67	2.77
1000 ~ 1500	2.84	2.91	3.00	3.11

Example 1.5

The specific weight of sea water is 1.026. Determine the mass density of sea water on the ocean floor 2000 m deep.

Solution

The change of pressure at the depth of 2000 m from that at the water surface is

$$\Delta P \cong (2000 \text{ m}) \times (1.026) \times (9810 \text{ N/m}^3)$$
$$= 20,130,000 \text{ N/m}^2$$

From Eq. (1.4) we have,

$$\Delta P = -E_b\left(\frac{\Delta\text{Vol}}{\text{Vol}_0}\right)$$

where $E_b = 2.34 \times 10^9$ (approximately at 20°C).

$$\frac{\Delta\text{Vol}}{\text{Vol}_0} = \frac{-20,130,000}{2.34 \times 10^9} = -0.0086$$

As density is defined as mass per unit volume,

$$\rho_0 = \frac{m}{\text{Vol}_0} \quad \text{and} \quad \rho = \frac{m}{\text{Vol}}$$

Then

$$\Delta\text{Vol} = \frac{m}{\rho} - \frac{m}{\rho_0} \quad \text{and} \quad \frac{\Delta\text{Vol}}{\text{Vol}_0} = \frac{\rho_0}{\rho} - 1$$

So that

$$\rho = \frac{\rho_0}{\left(1 + \dfrac{\Delta \text{Vol}}{\text{Vol}_0}\right)} = \frac{1.026}{1 - 0.0086} = 1.035$$

PROBLEMS

1.5.1. Determine the change in density of water at 20°C when pressure is suddenly increased from 25 bar to 4,500,000 dyn/cm².

1.5.2. The volume of a liquid decreased by 0.8% due to the increase of pressure from 5000 N/cm² to 10,000 N/cm². What is the bulk modulus of elasticity of the liquid?

REVIEWING PROBLEMS

1.1 A steel tank 120 m³ in volume is full of water at atmospheric pressure. Find the weight of the water at 4°C. Compare the weight of water at the pressure of 10 bar, and at 100 bar.

1.2 Establish the equivalence in absolute viscosity between 1 poise and 1 lb-sec/ft². Similarly, for the kinematic viscosity, find the equivalence between the stoke and the English unit.

1.3 The absolute viscosity of mercury is 1.7 centipoise and the specific gravity is 13.6 at standard pressure and temperature. Find the kinematic viscosity in English units.

1.4 At about 15.5°C, mercury has the same absolute viscosity as water. Find the corresponding kinematic viscosity.

1.5 A capillary tube is bent over the edge of a tank filled with water, as shown in Figure 1.7. (a) Discuss whether the surface tension on the tube could cause the water to flow over the edge of the tank. (b) With any choice of liquid, capillary tube diameter, and the level of liquid elevation in the tank, can the liquid drop out of the tube? Why?

Figure 1.7

2

WATER PRESSURE
AND PRESSURE FORCES

2.1 FREE SURFACE OF WATER

When water fills a containing vessel, it automatically seeks a horizontal surface upon which the pressure is constant everywhere. In practice, the free surface of water in a vessel is the surface that is not in contact with the cover of the vessel. Such a surface may be subjected to the atmospheric pressure (open vessel) or any other pressure that is exerted in the vessel (closed vessel).

2.2 ATMOSPHERIC PRESSURE
AND PRESSURE GAGES

The earth's atmosphere is a thick layer (approximately 1500 km) of mixed gases. Nitrogen makes up approximately 78 % of the atmosphere, oxygen makes up approximately 21 %, and the remaining 1 % consists mainly of argon and trace amounts of other gases. Since each of the gases possesses mass and consequently has weight, this weight exerts a pressure on every surface with which it comes in contact. At sea level, under normal conditions, the atmospheric pressure is approximately equal to that of a 76-cm-high mercury column or 10.33 m of water column. This is approximately equivalent to an absolute pressure of 10^5 N/m², or 1 bar*. The pressure unit 1 N/m² is also known as 1 Pascal, named after French mathematician Blaise Pascal (1623–1662).

*1 atmosphere $= 760 \text{ mmHg} = 10.33 \text{ mH}_2\text{O}$
$= 1.013 \cdot 10^5 \text{ N/m}^2 = 1.013 \cdot 10^5 \text{ Pascals}$
$= 1.013 \text{ bars} = 14.7 \text{ pounds/in.}^2$

To determine the variation of pressure between any two points in water, we may consider two arbitrary points A and B along an arbitrary x-axis as shown in Figure 2.1. Consider that these points lie in the ends of a small prism of water having a cross-sectional area dA and a length L. P_a and P_b are the pressures at each end, where the cross-sectional areas are normal to the x-axis. Since the prism is at rest, all forces acting upon it must be in equilibrium in all directions. For the force components in the x-direction, we may write

$$F_x = P_a \, dA - P_b \, dA - \gamma L \, dA \sin \theta = 0$$

Here $L \sin \theta = h$ is the vertical elevation difference between the two points. The above equation reduces to

$$P_a - P_b = \gamma h \tag{2.1}$$

Therefore, the *difference in pressure between any two points in still water is always equal to the product of the unit weight of water and the difference in elevation between the two points.*

If the two points are on the same elevation, $h = 0$ and $P_a = P_b$. In other words, *for water at rest, the pressure at all points in a horizontal plane is the same.* If the water body has a free surface that is exposed to *atmospheric pressure*, P_{atm}, we may position point B on the free surface and write

$$(P_a)_{\text{abs}} = \gamma h + P_b = \gamma h + P_{\text{atm}} \tag{2.2}$$

This pressure, $(P_a)_{\text{abs}}$, is commonly referred to as the *absolute pressure.*

Pressure gages are usually designed to measure pressures above or below the atmospheric pressure. Pressure so measured, using atmospheric pressure as a base, is called *gage pressure*, P. Absolute pressure is always equal to gage pressure plus atmospheric pressure:

$$(P_a)_{\text{abs}} = P + P_{\text{atm}} \tag{2.3}$$

Figure 2.2 shows two typical pressure gage dials. Comparing Eqs. (2.2) and

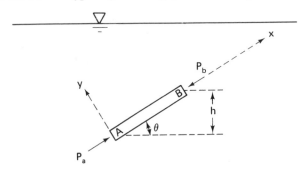

Figure 2.1. Hydrostatic pressure on a prism.

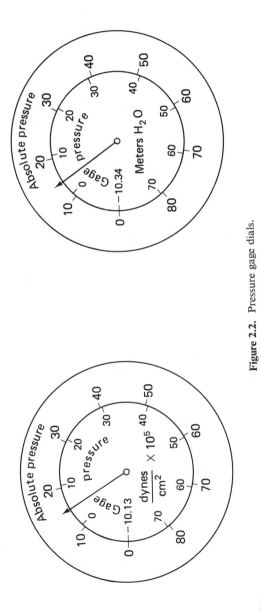

Figure 2.2. Pressure gage dials.

15

(2.3), we have

$$P = \gamma h \tag{2.4}$$

or

$$h = \frac{P}{\gamma} \tag{2.5}$$

Here the pressure is expressed in terms of the height of a water column h. In hydraulics it is known as the *pressure head*.

Equation (2.1) may thus be rewritten as

$$\frac{P_a}{\gamma} - \frac{P_b}{\gamma} = \Delta h \tag{2.6}$$

meaning that the difference in pressure heads at two points in water at rest is always equal to the difference in elevations between the two points. From this relationship we can also see that any change in pressure at point B would cause an equal change at point A, for the difference in pressure head between the two points must remain the same value h. In other words, *a pressure applied at any point in a liquid at rest is transmitted equally and undiminished in all directions to every other point in the liquid.* This principle, also known as *Pascal's law*, has been made use of in the hydraulic jacks that lift heavy weights by applying relatively small forces.

Example 2.1

The diameters of cylindrical pistons F and W are 3 cm and 20 cm, respectively. The faces of the pistons are at the same elevation and the intervening passages are filled with an incompressible hydraulic oil. A force P of 100 N is applied at the end of the lever, as shown in Figure 2.3. What weight W can the hydraulic jack support?

Measurements in cm

Figure 2.3. Hydraulic jack.

Solution

Balancing the moments produced by P and F, we obtain for the equilibrium condition,

$$P \cdot (80 + 20) = F \cdot (20)$$
$$100 \cdot (100) = F \cdot (20)$$

Thus

$$F = 500 \text{ N}$$

From Pascal's law, the pressure P_a applied at A should be the same as that of P_b applied at B. We may write

$$P_a = \frac{F}{\frac{1}{4}(\pi \cdot 3^2)}; \qquad P_b = \frac{W}{\frac{1}{4}(\pi \cdot 20^2)}$$

$$\frac{500}{\frac{1}{4}(\pi \cdot 3^2)} = \frac{W}{\frac{1}{4}(\pi \cdot 20^2)}$$

$$\therefore \quad W = 500 \times \frac{20^2}{3^2} = 22{,}222 \text{ N}$$

PROBLEMS

2.2.1. A closed tank contains a liquid (sp. gr. 0.80) under pressure. The pressure gage, shown in Figure 2.4, indicates a pressure of 3.2×10^5 dyn/cm². Determine the pressure at the bottom of the tank and the height of the liquid column that will rise in the vertical tube.

Figure 2.4

2.2.2. An underwater storage tank was constructed offshore to store natural gas. Determine the gas pressure in the tank (in bars) when the water elevation in the tank is 6 m below the sea level (see Figure 2.5). The specific gravity of sea water is 1.025.

Figure 2.5

2.2.3. A certain freshwater fish does not survive well if it gets in water in which the absolute pressure is greater than 5 standard atmospheric pressure. How deep can the fish go before it experiences trouble?

2.2.4. A closed oil tank contains oil of specific gravity 0.85. If the pressure in the air space at the top of the oil surface measures 1.20 standard atmospheric pressure, what is the pressure at a point 5 m below the oil surface?

2.3   MANOMETERS

A manometer is usually a tube bent in the form of a U containing a fluid of known specific gravity. The difference in elevations of the liquid surfaces, under pressure, indicates the difference in pressure at the two ends. Basically, there are two types of manometer: (1) An *open manometer* has one end open to atmospheric pressure and is capable of measuring the gage pressure in a vessel. (2) A *differential manometer* connects each end to a different pressure vessel and is capable of measuring the pressure difference between the two vessels.

The liquid used in a manometer is usually heavier than the fluids to be measured. It must also form a meniscus, that is, it must not mix with the adjacent liquids. The most frequently used manometer liquids are mercury (sp. gr. 13.6), water (sp. gr. 1.00), alcohol (sp. gr. 0.9), and other commercial manometer oils of various specific gravities (e.g., Meriam* Unit Oil, sp. gr. 1.00; Meriam No. 3 Oil, sp. gr. 2.95; etc).

Figure 2.6 shows the schematics of a typical open manometer. It is obvious that the higher the pressure in vessel A, the larger the difference, h, in the surface elevations in the two legs of the manometer. The accurate calculation of pressure in A, however, must involve the consideration of the densities of the fluids and the geometry involved in the entire measuring system.

A simple step-by-step procedure is suggested for manometer computation:

(Step 1) Make a sketch of the manometer system, similar to the one in Figure 2.6, approximately to scale.

(Step 2) Draw a horizontal line at the level of the lower surface of the manometer liquid, 1. The pressures at points 1 and 2 must be the same since the system is in static equilibrium.

*Meriam Instruments, Division of Scott & Fetzer Co., Cleveland, Ohio 44102.

Figure 2.6. Open manometer.

(Step 3) The pressure on 2 is exerted by the weight of the liquid column above 2; and the pressure on 1 is equal to the weight of the column of liquid A above 1 plus the pressure in vessel A. They must be equal in value. This relation may be written as follows:

$$\gamma_B h = \gamma_A y + P_A \quad \text{or} \quad \frac{\gamma_B}{\gamma_A} h = y + \frac{P_A}{\gamma_A}$$

Either one of these equations can be used to solve for P_A. The same procedure can be applied to any complex geometry, as shown in the following example.

Example 2.2

Mercury is used as the manometer liquid to measure the water pressure in vessel A. Determine the pressure when the manometer readings are as shown in Figure 2.7.

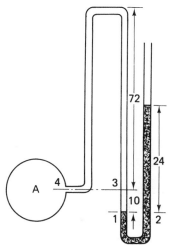

Measurements in cm

Figure 2.7

Solution

(Step 1) is accomplished by the horizontal reference lines in Figure 2.7.
   (Step 2) Since the system is in equilibrium, we may write that

$$P_1 = P_2 \tag{a}$$

and

$$P_3 = P_A \tag{b}$$

(Step 3) To balance the pressure at the level of 1-2, we have

$$13.6 \cdot (24) \text{ cm} = 1.0 \cdot (10) \text{ cm} + \frac{P_3}{981 \text{ dyn/cm}^3}$$

Thus

$$P_3 = (326.4 - 10) \cdot 981 = 310,380 \text{ dyn/cm}^2 = P_A$$

   The open manometer in a U-tube, as described above, requires readings
of liquid levels at two points. Since any change in pressure in the vessel causes
a drop of liquid surface at one end and a rise in the other, a *single-reading
manometer* can be made by introducing a reservoir with a large cross-sectional area, compared to that of the tube, into one leg of the manometer. A
typical single-reading manometer is shown in Figure 2.8.

Figure 2.8. Single-reading manometer.

   Due to the large area ratio between the reservoir and the tube, a small
drop of surface elevation in the reservoir will cause an appreciable rise of
liquid column in the other leg. If there is an increase of pressure, ΔP_A will
cause the liquid surface in the reservoir to drop by a small amount Δy. Then

$$A \cdot \Delta y = a \cdot h \tag{2.7}$$

where A and a are cross-sectional areas of the reservoir and the tube, respectively.

Applying (step 2) to points 1 and 2, we may write generally

$$\gamma_A \cdot (y + \Delta y) + P_A = \gamma_B \cdot (h) \tag{2.8}$$

Simultaneous solution of Eqs. (2.7) and (2.8) gives the value of P_A, the pressure in the vessel, in terms of h. All other quantities in Eqs. (2.7) and (2.8), A/a, y, γ_A, and γ_B, are quantities predetermined in the manometer design. A single reading of h will thus determine the pressure.

Since Δy can be made negligible by introducing a very large A/a ratio, the above relationship may be further simplified to

$$\gamma_A \cdot y + P_A = \gamma_B \cdot h \tag{2.9}$$

The height reading h is the measure of the pressure in the vessel.

The solution of practical hydraulic problems frequently requires the difference in pressure between two points in a pipe or a pipe system. For this purpose, differential manometers are frequently used. A typical differential manometer is shown in Figure 2.9.

The same computation steps suggested on pages 18 and 19 can be readily applied here too: When the system is in static equilibrium, the pressure at the same elevation, points 1 and 2, must be equal. We may thus write

$$\gamma_A \cdot (y + h) + P_A = \gamma_B \cdot h + \gamma_A \cdot y + P_B$$

Hence, the pressure difference, P, is expressed as

$$P = P_B - P_A = (\gamma_A - \gamma_B) \cdot h \tag{2.10}$$

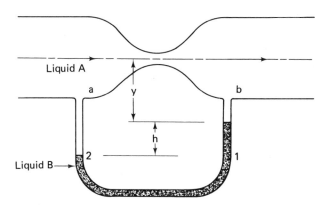

Figure 2.9. A differential manometer installed in a flow measurement system.

PROBLEMS

2.3.1. A U-tube with both ends open to the atmosphere contains mercury in the lower portion of the tube. If water is filled into one leg of the U-tube until the water column is 1 m above the mercury-water meniscus, what is the elevation difference between the mercury surfaces in the two legs?

2.3.2. If oil (sp. gr. 0.79) is filled into the other leg of the above manometer to a height of 60 cm, what is the elevation difference between the mercury surfaces?

2.3.3. A manometer is mounted on a city water supply main pipe to monitor the water pressure in the pipe, as shown in Figure 2.10. Determine the water pressure.

Figure 2.10

2.3.4. An open manometer, shown in Figure 2.11, is installed to measure the pressure in a pipe carrying an oil (sp. gr. 0.82). If the manometer liquid is carbon tetrachloride (sp. gr. 1.60), determine the pressure in the pipe (in meters of water column height).

Figure 2.11

2.3.5. In Figure 2.12 a single-reading mercury manometer is used to measure water pressure in the pipe. What is the pressure if $h_1 = 20$ cm and $h_2 = 67$ cm?

Figure 2.12

2.3.6. In Figure 2.13 fluids A and B have specific gravities of 0.75 and 1.00, respectively. If mercury is used as the manometer liquid, determine the pressure difference between A and B.

Figure 2.13

2.4 HYDROSTATIC FORCE ON SURFACES

Let us take an arbitrary area AB on the backface of a dam that inclines at an angle θ and then place the x-axis on the line at which the water-free surface intersects with the dam surface, with the y-axis running down the direction of the dam surface. Figure 2.14(a) shows a horizontal view of the area and Figure 2.14(b) shows the projection of AB on the dam surface.

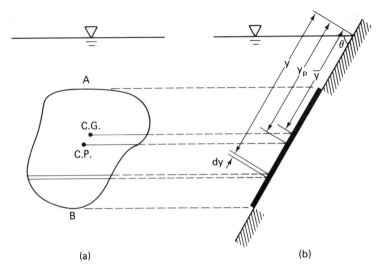

Figure 2.14. Hydrostatic pressure on a plane surface.

We may assume that the plane surface AB is made up of an infinite number of horizontal strips, each having a width of dy and an area of dA. The hydrostatic pressure on each strip may be considered constant because the width of each strip is very small. For a strip at depth h below the free surface, the pressure is

$$p = \gamma h = \gamma y \sin \theta$$

The total pressure force on the strip is the pressure times the area:

$$dP = \gamma y \sin \theta \, dA$$

and the total pressure force over the entire AB plane surface is the sum of pressure on all the strips:

$$P = \int_A dP = \int_A \gamma \cdot y \cdot \sin \theta \cdot dA = \gamma \cdot \sin \theta \cdot \int_A y \, dA \qquad (2.11)$$
$$= \gamma \cdot \sin \theta \cdot A \cdot \bar{y}$$

where $\bar{y} = \int_A y \, dA / A$ is the distance measured from the x-axis to the centroid (c.g.), of the AB plane.

Substituting $\bar{h}$, the vertical distance of the centroid below the water surface, for $\bar{y} \sin \theta$, we have

$$P = \gamma \cdot \bar{h} \cdot A \qquad (2.12)$$

This equation states the fact that *the total hydrostatic pressure force on any submerged plane surface is equal to the product of the surface area and the pressure acting at the centroid of the plane surface.*

Pressure forces acting on a plane surface are parallel and distributed over every part of the surface. These parallel forces can be analytically replaced by a single resultant force P of the same magnitude and direction as shown in Eq. (2.12). The point on the plane surface at which this resultant force acts is known as the *center of pressure.* Considering the plane surface as a free body, we see that the distributed forces can be replaced by the single resultant force at the pressure center without altering any reactions or moments in the system. Designating Y_p as the distance measured from the x-axis to the center of pressure, we may thus write

$$P \cdot Y_p = \int_A y \cdot dP$$

Hence

$$Y_p = \frac{\int_A y \, dP}{P} \tag{2.13}$$

Substituting the relationships $dP = \gamma y \sin\theta \cdot dA$ and $P = \gamma \sin\theta \cdot A \cdot \bar{y}$, we may write Eq. (2.13) as

$$Y_p = \frac{\int_A y^2 \, dA}{A\bar{y}} \tag{2.14}$$

in which $\int_A y^2 \, dA = I_x$ and $A \cdot \bar{y} = M_x$ are, respectively, the moment of inertia and the static moment of the plane surface AB with respect to the x-axis. Therefore,

$$Y_p = \frac{I_x}{M_x} \tag{2.15}$$

With respect to the centroid of the plane, they may be written as

$$Y_p = \frac{I_0 + A\bar{y}^2}{A\bar{y}} = \frac{I_0}{A\bar{y}} + \bar{y} \tag{2.16}$$

where I_0 is the moment of inertia of the plane with respect to its own centroid, A is the plane surface area, and $\bar{y}$ is the distance between the centroid and the x-axis.

The center of pressure of any submerged plane surface is always below the centroid of the surface. All three quantities in the first term on the right-hand side of Eq. (2.16) are positive quantities.

The centroid, area, and moment of inertia with respect to the centroid of certain common geometrical plane surfaces are given in Table 2.1.

TABLE 2.1 Surface Area, Center of Gravity, and Moment of Inertia of Certain Simple Geometrical Plates

Shape	Area	Centroid	Moment of inertia about the neutral axis
Rectangle	$b \cdot h$	$\bar{x} = \frac{1}{2} b$ $\bar{y} = \frac{1}{2} h$	$I_0 = \frac{1}{12} b \cdot h^3$
Triangle	$\frac{1}{2} b \cdot h$	$\bar{x} = \frac{b + c}{3}$ $\bar{y} = h/3$	$I_0 = \frac{1}{36} b \cdot h^3$
Circle	$\frac{1}{4} \pi d^2$	$\bar{x} = \frac{1}{2} d$ $\bar{y} = \frac{1}{2} d$	$I_0 = \frac{1}{64} \pi d^4$
Trapezoid	$\frac{h(a + b)}{2}$	$\bar{y} = \frac{h(2a + b)}{3(a + b)}$	$I_0 = \frac{h^3(a^2 + 4ab + b^2)}{36(a + b)}$
Semicircle	$\frac{1}{2} \pi r^2$	$\bar{y} = \frac{4r}{3\pi}$	$I_0 = \frac{(9\pi^2 - 64) r^4}{72 \pi}$

TABLE 2.1 Cont.

Shape	Area	Centroid	Moment of inertia about the neutral axis
Ellipse			

| | πbh | $\overline{x} = b$ | $I_0 = \dfrac{\pi}{4} b \cdot h^3$ |
| | | $\overline{y} = h$ | |

Semi-ellipse

| | $\dfrac{\pi}{2} b \cdot h$ | $\overline{x} = b$ | $I_0 = \dfrac{(9\pi^2 - 64)}{72\pi}$ |
| | | $\overline{y} = \dfrac{4h}{3\pi}$ | |

Parabolic section

| $y = f(x)$ | $\dfrac{2}{3} b \cdot h$ | $\overline{y} = \dfrac{2}{5} h$ | |
| $y = h(1 - \dfrac{x^2}{b^2})$ | | $\overline{x} = \dfrac{3}{8} b$ | $I_0 = \dfrac{8}{175} b \cdot h^3$ |

Example 2.3

A vertical trapezoidal gate with its upper edge located 5 m below the free surface of water is shown in Figure 2.15. Determine the total pressure force and the center of pressure on the gate.

Solution

The total pressure force is

$$P = \gamma \cdot \bar{h} \cdot A$$

$$= 9810 \cdot \left[5 + \frac{2 \cdot (2 \cdot 1 + 3)}{3 \cdot (1 + 3)} \right] \cdot \left[\frac{2 \cdot (3 + 1)}{2} \right]$$

$$= 228,900 \text{ N}$$

The location of the center of pressure is, (Eq. 2.16),

$$Y_P = \frac{I_0}{A\bar{y}} + \bar{y}$$

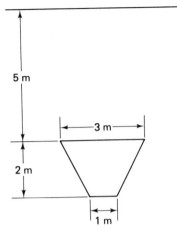

Figure 2.15

where

$$I_0 = \frac{2^3 \cdot [1^2 + 4(1)(3) + 3^2]}{36 \cdot (1 + 3)} = 1.22 \text{ m}^4$$

$$\bar{y} = 5.83 \text{ m}$$

$$A = 4 \text{ m}^2$$

Thus

$$Y_p = \frac{1.22}{4 \cdot (5.83)} + 5.83 = 5.88 \text{ m}$$

below the water surface.

Example 2.4

An inverted semicircle gate (see Figure 2.16) is installed at 45° with respect to the free water surface. The top of the gate is 5 m below the water surface. Determine the total pressure and the center of pressure on the gate.

Solution

From Eq. (2.11)

$$P = \gamma \cdot \sin \theta \cdot A \cdot \bar{y}$$

where

$$A = \frac{1}{2}\left(\frac{\pi}{4} \cdot 4^2\right) = 6.28 \text{ m}^2$$

and

$$\bar{y} = 5 \sec 45° + \frac{4 \cdot 4}{3\pi} = 8.77 \text{ m}$$

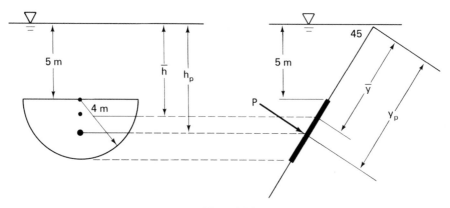

Figure 2.16

Therefore, $P = 9810 \cdot (\sin 45°)(6.28)(8.77) = 382,044$ N. This is the total hydrostatic force active on the gate. From Eq. (2.16)

$$Y_P = \frac{I_0}{A\bar{y}} + \bar{y}$$

From Table 2.1

$$I_0 = \frac{\pi h^2}{2}$$

$$= \frac{\pi (4)^2}{2} = 25.13 \text{ m}^4$$

Therefore,

$$Y_P = \frac{25.13}{6.28(8.77)} + 8.77 = 9.23 \text{ m}$$

This is the distance measured from the top of the water surface to the pressure center of the gate.

Pressure force on a curved surface can be best analyzed by resolving the total pressure force on the surface's horizontal and vertical components. Figure 2.17 shows a curved wall of a container vessel having a unit thickness normal to the plane of the paper.

Since the water body in the container is stationary, every part of the water body must be in equilibrium, or, the force components must each satisfy the equilibrium conditions, i.e., $\Sigma F_x = 0$ and $\Sigma F_y = 0$.

In the free body of water ABA', where $A'B$ is the vertical projection of AB, we can clearly see that the horizontal pressure component P_h must equal the horizontal pressure exerted on plane surface $A'B$. The vertical pressure component, P_v, must equal the weight of the water body contained within

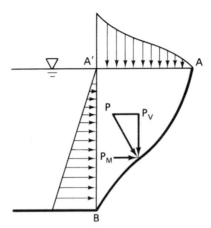

Figure 2.17. Hydrostatic pressure on a curved surface.

volume ABA', since the vertical component of the pressure force exerted on plane surface $A'B$ is zero and the plane cannot take any shear force. In other words, it may be stated that

(1) *The horizontal component of the total hydrostatic pressure force on any surface is always equal to the total pressure on the vertical projection of the surface. The resultant force of the horizontal component can be located through the center of pressure of this projection.*

(2) *The vertical component of the total hydrostatic pressure force on any surface is always equal to the weight of the entire water column above the surface extending vertically to the free surface. The resultant force of the vertical component can be located through the centroid of this column.*

Example 2.5

Determine the total hydrostatic pressure and the center of pressure on the 3-m-long, 2-m-high quadrant gate in Figure 2.18.

Solution

The horizontal component is equal to the hydrostatic pressure force on the projection plane $A'B$.

$$P_h = \gamma \cdot \bar{h} \cdot A = 9810\left(\frac{1}{2} \cdot 2 \cdot 2 \cdot 3\right) = 58,860 \text{ N}$$

The pressure center of the horizontal component is located at $2/3 \times 2 = 1.33$ m below the free surface. The vertical component is equal to the weight of

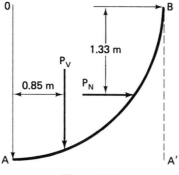

Figure 2.18

water replaced by the volume AOB. The direction of this pressure component is upward, since the water volume is replaced.

$$P_v = \gamma V = 9810\left[\frac{1}{4}\pi\cdot(2^2)\right](3) = 92{,}460 \text{ N}$$

The pressure center is located at $4(2)/3\pi = 0.85$ m. The resultant force of water on AB is

$$P = \sqrt{(58{,}860)^2 + (92{,}460)^2} = 109{,}600 \text{ N}$$

$$\theta = \tan^{-1}\left(\frac{P_v}{P_h}\right) = \tan^{-1}\frac{92{,}460}{58{,}860} = 57.5°$$

PROBLEMS

2.4.1. The rectangular gate in Figure 2.19 is hinged at A and separates water in the reservoir from the tail water tunnel. If the uniform thickness gate has a dimension of 2 m × 3 m and weighs 2 tons, determine the maximum height h for which the gate will stay closed.

Figure 2.19

2.4.2. A circular gate is installed on a vertical wall as shown in Figure 2.20. Determine the horizontal force, $\bar{T}$, necessary in order to hold the gate at the edge (in terms of diameter, D, and height, h). Neglect friction at the pivot.

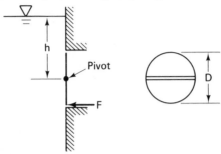

Figure 2.20

2.4.3. In Figure 2.21 the wicket dam is 5 m high and 2 m wide and is pivoted at its center. Determine the reaction force in the supporting member AB.

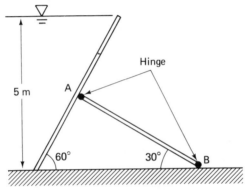

Figure 2.21

2.4.4. The tainter gate section shown in Figure 2.22 has a cylindrical surface with a 10-m radius and is supported by a structural frame hinged at O. The gate is 12 m long (in the direction perpendicular to the paper). Determine the magnitude and location of the total hydrostatic force on the gate (*Hint*: Horizontal and vertical force components).

Figure 2.22

2.4.5. In Figure 2.23 the corner plate of a large hull is curved with a radius of 1.75 m. When the barge is submerged in sea water (sp. gr. 1.025), determine the magnitude and location of pressure on the curved plate *AB* per unit length of the hull.

Figure 2.23

2.4.6. The cylindrical dome in Figure 2.24 is 10 m long and is secured to the top of an oil tank by bolts. If the oil has a specific gravity of 0.86 and the pressure gage reads 2.42 bar, determine the total tension force in the bolts. Neglect the weight of the cover.

Figure 2.24

2.5 BUOYANCY AND FLOTATION STABILITY

Archimedes (about 250 B.C.) discovered that *the weight of a submerged body is reduced by an amount equal to the weight of the liquid displaced by the body.* *Archimedes' principle*, as we now call it, can be easily proven by using the formula described in Section 2.4.

Assume that a solid body of arbitrary shape, *AB*, is submerged in water as shown in Figure 2.25. A vertical plane *MN* may then be drawn through the body in the direction normal to the paper. One observes that the horizontal pressure force component in the direction of the paper, P_h and P'_h, must be equal because they both are measured by the same vertical projection area *MN*. The horizontal pressure force component in the direction normal to the

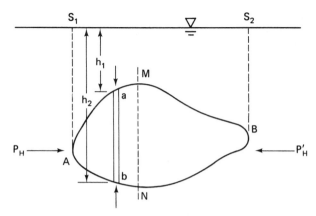

Figure 2.25. Buoyancy of a submerged body.

paper must also be equal for the same reason because they share the projection in the plane of the paper.

The vertical pressure force component can be analyzed by taking a small vertical prism ab with a cross-sectional area ΔA. The vertical pressure force on top of the prism is $(\gamma h_1 \cdot dA)$ acting downward. The vertical force on the bottom of the prism is $(\gamma h_2 \cdot dA)$ acting upward. The difference gives the resultant vertical force component on the prism:

$$P_v = \gamma h_2 \cdot dA - \gamma h_1 \cdot dA = \gamma(h_2 - h_1)\, dA \qquad \text{(upward)}$$

which is exactly equal to the weight of water column ab replaced by the prism. In other words, the weight of the submerged prism is reduced by an amount equal to the weight of the liquid replaced by the prism. A summation of the vertical forces on all the prisms that make up the entire submerged body AB gives the proof of Archimedes' principle.

Archimedes' principle may also be looked upon as an action of the difference of vertical pressure forces on the two surfaces ANB and AMB. The vertical pressure force on surface ANB is equal to the weight of the hypothetical water column (volume of S_1ANBS_2) acting upward; and the vertical pressure force on surface AMB is equal to the weight of the water column S_1AMBS_2, acting downward. Since the volume S_1ANBS_2 is larger than the volume S_1AMBS_2 by an amount exactly equal to the volume of the submerged body $AMBN$, the net difference is a force equal to the weight of water that would be contained in the volume $AMBN$ acting upward. This force is the buoyancy on the body.

A floating body is a body partially submerged due to the balance of the body weight and the buoyancy force.

Example 2.6

The barge in Figure 2.26 is 12 m long, 2 m high, and 3 m wide, and weighs 200,000 N. Determine the minimum depth of the waterway necessary to carry the barge.

Figure 2.26

Solution

As the barge floats, the weight of the water displaced by the barge equals the weight of the barge. Therefore, the minimum depth of the waterway is the submerged depth of the barge, x, which may be calculated as

$$\text{weight of barge} = \text{weight of water replaced by the submerged}$$
$$\text{portion of barge}$$
$$200,000 = 9810(12) \cdot (3) \cdot (x)$$
$$(x) = 0.566 \text{ m}$$

The minimum depth of the waterway necessary to carry the barge would have to be 0.566 m.

The stability of a floating body is determined by the relative positions of the center of gravity of body G and the *center of buoyancy B* which is the center of gravity of the liquid volume replaced by the body, as shown in Figure 2.27.

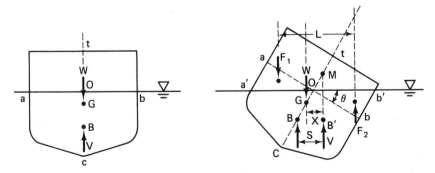

Figure 2.27. Center of buoyancy and metacenter of a floating body.

The body is in equilibrium if its center of gravity and its center of buoyancy lie on the same vertical line, as in Figure 2.27(a). This equilibrium may be disturbed by a variety of causes, for example, wind or wave action. The floating body is made to heel through an angle θ as shown in Figure 2.27(b). When the floating body is in the heeled position, the center of gravity of the body remains unchanged but the center of buoyancy, which is now the center of gravity of area $a'cb'$, has been changed from B to B'. The buoyancy force V, acting upward through B', and the weight of the body W, acting downward through G, constitute a couple, $W \cdot x$, which resists further overturning and tends to restore the body to its original equilibrium position.

By extending the line of vertical buoyancy force action through the buoyancy B', we see that the vertical line intersects the axis of orginal symmetry c–t at a point M. The point M is known as the *metacenter* of the floating body, and the distance between the center of gravity and the metacenter is known as the *metacentric height*. The metacentric height is a measure of the flotation stability of the body. When the angle of inclination is small, the position of M does not change materially with the tilting position. The metacentric height and its righting moment can be determined in the following way.

Since the tilting of a floating body does not change the total weight of the body, the total displacement volume is not changed. The roll through an angle θ only changes the shape of the displaced volume due to the adding of the immersion wedge bob' and the subtracting of the emersion wedge aoa'. In this new position, the total buoyancy force γVol is shifted through a horizontal distance S to B', and it may be considered as the result of compounding the force γVol in its original position at B and a couple F_1 and F_2, which caused γVol to change position. The moment of the resultant force about point B must equal the algebraic sum of the moments about its components:

$$\gamma \cdot \text{Vol} \cdot S = \gamma \cdot \text{Vol} \cdot (\text{zero}) + \text{moment of the force couple}$$
$$= 0 + \gamma \cdot \text{vol} \cdot L$$

or

$$\gamma \cdot \text{Vol} \cdot S = \gamma \cdot \text{vol} \cdot L$$
$$S = \frac{\text{vol}}{\text{Vol}} \cdot L \tag{a}$$

where vol is the volume of wedge bob' (or aoa') and L is the horizontal distance between the centers of gravity of the two wedges.

But, according to the geometric relation, we have

$$S = \overline{MB} \cdot \sin \theta \quad \text{or} \quad \overline{MB} = \frac{S}{\sin \theta} \tag{b}$$

Comparing Eqs. (a) and (b), we get

$$\overline{MB} = \frac{\text{vol} \cdot L}{\text{Vol} \cdot \sin \theta}$$

For a small angle, $\sin \theta \simeq \theta$, the above relationship may be simplified:

$$\overline{MB} = \frac{\text{vol} \cdot L}{\text{Vol} \cdot \theta}$$

The buoyancy force produced by wedge *bob'* (see Figure 2.28) can be estimated by considering a small prism of the wedge. Assume that the prism has a horizontal area, *dA*, and is located at a distance *x* from the axis of rotation, *O*. The height of the prism is $x \cdot (\tan \theta)$. For small angle θ, it may be approximated by $x \cdot \theta$. Thus, the buoyancy force produced by this small prism is $\gamma x \cdot \theta \, dA$. The sum of the moment produced by it, about the axis of rotation *O*, is $\gamma x^2 \theta \, dA$. The sum of the moment produced by each of the prisms in the wedge gives the moment of the immersed wedge. The moment produced by the force couple is, therefore,

$$F \cdot L = \int_A \gamma x^2 \, \theta \, dA = \gamma \theta \int_A x^2 \, dA$$

Since the moment of the couple must be equal to $\gamma \cdot \text{vol} \cdot L$, and $S = \overline{MB} \cdot \theta$ for small θ, we may write

$$\int_A x^2 \, dA = \overline{MB} \cdot \text{Vol}$$

But $\int_A x^2 \, dA$ is the moment of inertia of the waterline cross-section area of the floating body about the axis of rotation *O*.

$$I_o = \int_A x^2 \, dA$$

For small angles of tilting, the moment of inertia for upright cross section *aob* and the inclined cross section *a'ob'* may be approximated by a constant

Figure 2.28

value. Therefore,

$$\overline{MB} = \frac{I_o}{\text{Vol}} \tag{2.17}$$

The metacentric height, defined as the distance between the metacenter M and the center of gravity G can be estimated:

$$\overline{GM} = \overline{MB} \pm \overline{GB} = \frac{I_o}{\text{Vol}} \pm \overline{GB} \tag{2.18}$$

The distance between the center of gravity and the center of buoyancy $\overline{GB}$ in the upright position, shown in Figure 2.27, can be determined by the sectional geometry or the design data of the vessel.

The ($\pm$) sign, taken as the center of gravity of a floating body, may be located above or below the center of buoyancy. For greater stability of flotation, it is obviously advantageous to have a negative value for $\overline{GB}$ or to make the center of gravity as low as possible.

The *righting moment*, when tilted, is

$$M = W \cdot \overline{GM} \cdot \sin \theta \tag{2.19}$$

The stability of buoyant bodies under various conditions may be summarized as follows:

1. A floating body is stable if the center of gravity is below the metacenter. Otherwise, it is unstable.

2. A submerged body is stable if the center of gravity is below the center of buoyancy.

Example 2.7

A 3 m $\times$ 4 m rectangular box caisson is 2 m deep (see Figure 2.29). It has a draft of 1.2 m when floating in an upright position. Compute (a) the metacentric height and (b) the righting moment in sea water (sp. gr. 1.03) when the angle of heel is 8°.

Solution

From Eq. 2.18

$$\overline{GM} = \overline{MB} - \overline{GB}$$

where

$$\overline{MB} = \frac{I_o}{\text{Vol}}$$

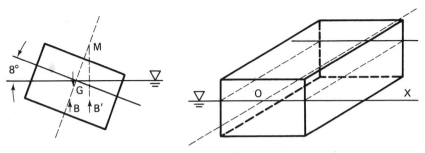

Figure 2.29

and I_o is the waterline area moment of inertia of the box about its longitudinal axis through O.

Then

$$\overline{GM} = \frac{\frac{1}{12}b^3 l}{b \cdot l \cdot 1.2} - \left(\frac{h}{2} - \frac{1.2}{2}\right)$$

$$= 0.225 \text{ m}$$

The specific gravity of sea water is 1.03, and from Eq. 2.19, the right moment is

$$M = W \cdot \overline{GM} \sin \theta$$

$$= 9810 \cdot 1.03 \cdot (4 \cdot 3 \cdot 1.2) \cdot (0.225)(\sin 8°)$$

$$= 4560 \text{ N-m}$$

PROBLEMS

2.5.1. A solid brass sphere (see Figure 2.30) of 30 cm diameter is used to weight down a cylindrical buoy in sea water (sp. gr. 1.03). The buoy has a height of 2 m and a specific gravity of 0.5, and is tied to the sphere at one end. What rise in tide, h, will be required to lift the anchor off the bottom?

Figure 2.30

2.5.2. A concrete block that has a total volume of 1 m^3 and a specific gravity of 1.67 is tied to one end of a long, hollow cylinder (similar arrangement as in Figure 2.30). The cylinder is 3 m long and has a 72-cm diameter. In deep water, 12 cm of the cylinder remain above the water surface. Determine the weight of the cylinder.

2.5.3. A rectangular barge is 12 m long, 5 m wide, and 2 m deep. The center of gravity is 1.0 m from the bottom and the barge draws 1.5 m of water. Find the metacenter height and the distance measured from the bottom to the metacenter for the following angles of heel: 4°, 8°, and 12°. What are the respective righting moments?

2.5.4. A 10-m diameter, cylindrical caisson floats upright in the ocean (sp. gr. 1.03). The center of gravity measures 2 m from the bottom which is 8 m below the water surface. Determine the metacenter height and the righting moment when the caisson is tipped through an angle of 10°.

2.5.5. A wooden block is 1 m wide, 1 m deep, and 2 m long. What is the specific gravity of the block if the metacenter is at the same point as the center of gravity? Is the floating block stable? Explain.

2.5.6. A 4.8 m wide, 12 m long, and 4.2 m deep rectangular pontoon has a draft of 2 m in sea water (sp. gr. 1.03). If the maximum angle of heel is 15°, determine the load that can be moved from the center to the edge of the pontoon.

REVIEWING PROBLEMS

2.1 The gage pressure on an open-top water tank reads 30 mm of mercury. Determine the water depth above the gage. Find the equivalency in bar, N/m^2 in absolute pressure at 20°C.

2.2 An open tank contains water of height h and an equally tall layer of oil with specific gravity of 0.82. If the gage pressure at the bottom of the tank (referred to ambient atmosphere) is to be 16 mm of mercury, determine the height h.

2.3 A micromanometer consists of two large reservoirs and a U-tube, as shown in Figure 2.31. If the density of the two liquids are, respectively, ρ_1, and ρ_2, determine an expression for the pressure difference in terms of ρ_1, ρ_2, h, d_1, and d_2.

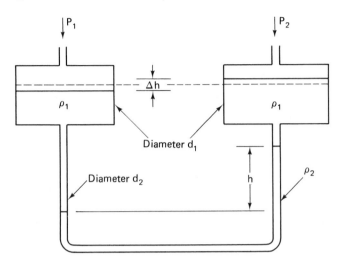

Figure 2.31

2.4 A concrete dam with a triangular cross-section as shown in Figure 2.32 is built to hold 10 m of water. The specific gravity of concrete is 1.67. Determine the moment generated respect to the toe of the dam, A.

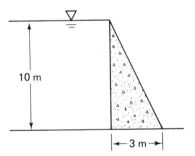

Figure 2.32

2.5 An inverted spherical shell of diameter d as shown in Figure 2.33 is used to cover a tank filled with water at 20°C. Determine the minimum weight of the shell needed to hold itself in place.

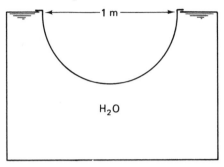

$$\text{—1 m—}$$

$$H_2O$$

Figure 2.33

2.6 Figure 2.34 shows a buoy which consists of a wooden pole 25 cm in diameter and 2 m long, with a spherical weight at the bottom. The specific weight of the wood is 0.62 and that of the bottom is 1.62. Determine the position of the center of buoyancy from the top of the buoy, and the metacenter height.

2 m

0.25 m

0.5 m

Figure 2.34

3

WATER FLOW
IN PIPES
AND PIPE NETWORKS

3.1 DESCRIPTION OF A PIPE FLOW

In hydraulics the term *pipe flow* usually refers to full water flow in closed conduits of circular cross sections under a certain pressure gradient. For a given discharge Q, the pipe flow at any section of the pipe can be described by the pipe cross section, the pipe elevation, the pressure, and the flow velocity in the pipe.

The elevation h of a particular section in the pipe is usually measured with respect to a horizontal reference datum, such as the *mean sea level* (MSL) at a certain estuary. The pressure in a pipe generally varies from one point to another, but a mean value is normally used over a particular section of interest. In other words, the regional pressure variation in a given cross section is commonly neglected unless otherwise specified. Figure 3.1 is a typical representation of a pipe flow from a large reservoir.

In most engineering computations the section mean velocity V, defined as the discharge, Q, divided by the cross-sectional area, A, is used.

$$V = \frac{Q}{A} \tag{3.1}$$

The velocity distribution within a cross section in a pipe, however, has special meaning in hydraulics. Its significance and importance in hydraulics are discussed next.

Figure 3.1

Near the end of the nineteenth century the French engineer Osborne Reynolds performed a very carefully prepared pipe flow experiment. Figure 3.2 shows the schematics of a typical setup for the Reynolds' experiment. A long, straight, glass pipe of small bore was installed in a large tank having glass sides. A cock C was installed at the outlet end of the glass pipe to regulate the outflow. A small bottle B filled with colored water and a regulating cock at the bottle's neck were used to introduce a fine stream of colored water into the entrance of the glass pipe when the flow was initiated. Water in the

Figure 3.2. Reynolds' apparatus.

large tank was allowed to settle very quietly in a room for several hours so that water in every part of the tank became totally stationary. Cock C was then opened partially to allow a very slow flow in the pipe. At this time the colored water appeared as a straight line extending to the downstream end, indicating *laminar flow* in the pipe. The flow rate was gradually increased until a certain velocity was reached, at which time the thread of color suddenly broke up and mixed with the surrounding water, showing that the pipe flow became turbulent at this point.

Reynolds found that the transition from laminar to turbulent flow in a pipe actually depends not only on the velocity, but also on the pipe diameter and the viscosity of the fluid. The relationship can be described by the ratio of the inertia force to the viscous force in the pipe. This ratio is commonly known as the *Reynolds number*, N_R, and it is shown (see also Chapter 8) in the following expression:

$$N_R = \frac{DV}{\nu} \qquad (3.2)$$

In the case of pipe flow, D is the pipe diameter, V is the mean velocity, and ν is the kinematic viscosity of the fluid, defined by the ratio of absolute viscosity μ and the fluid density ρ,

$$\nu = \frac{\mu}{\rho} \qquad (3.3)$$

It has been found and verified by many carefully prepared experiments that for flows in circular pipes the *critical Reynolds number* is about 2000. At this point the laminar pipe flow changes to a turbulent one. The transition from laminar to turbulent flow does not always happen at exactly $N_R = 2000$ but varies somewhat between one experiment and the other due to varying experimental conditions. This range of Reynolds number between laminar and turbulent flow is commonly known as the *transitional range* (see Figure 3.8).

Laminar flow occurs in a circular pipe when fluid flows in orderly laminae; this is analogous to the telescoping of a large number of thin-walled concentric tubes. The outer tube adheres to the pipe wall while the tube next to it moves with slightly higher velocity. The velocity of each successive tube increases gradually and reaches a maximum velocity near the center of the pipe. In this case, the velocity distribution takes the form of a paraboloid of revolution with the mean velocity V equal to one-half of the maximum centerline velocity, as shown in Figure 3.3.

In turbulent flow the turbulent motion causes the slower water particles adjacent to the pipe wall to mix continuously with the high-speed particles in the midstream; as a result, the high-speed particles in the midstream are slowed down due to the mixing. For this reason, the velocity distribution in

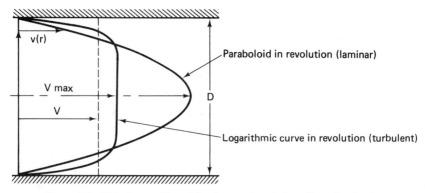

Figure 3.3. Velocity profiles of laminar and turbulent flows in circular pipes.

turbulent flow is more uniform (flatter) than that in laminar flow. The velocity profiles in turbulent pipe flows have been shown to take the general form of a logarithmic curve in revolution. Turbulent mixing activities increase with the Reynolds number; hence, the velocity distribution becomes flatter as the Reynolds number increases.

Under ordinary circumstances, water loses energy as it flows through a pipe. A major portion of the energy loss is caused by (1) friction against pipe walls and (2) viscous dissipation due to internal actions of the flow.

Wall friction on a moving column of water depends on the roughness of the wall material, e, and the velocity gradient, $\left.\dfrac{dV}{dr}\right|_{r=R_o}$ at the wall. For the same flow rate, it is seen (Figure 3.3) that turbulent flow has a higher wall velocity gradient than that of laminar flows; hence, a higher friction loss may be expected as the Reynolds number increases. At the same time, the internal mixing activities are also intensified as the flow becomes more turbulent, indicating an increasing rate of viscous dissipation in the flows. As a consequence, the rate of energy loss in pipe flow varies as a function of the Reynolds number and the roughness of the pipe wall. The engineering application of this consequence will be discussed later in this chapter.

3.2 ENERGY HEAD AND MAJOR HEAD LOSS

Water flow in pipes may contain energy in various forms. The major portion of the energy is contained in three basic forms: kinetic energy, potential energy, and pressure energy. The three forms of energy may be demonstrated by using a *stream tube*, as shown in Figure 3.4. A stream tube is a cylindrical passage with its surface everywhere parallel to the flow velocity; therefore, the flow cannot cross its surface.

The kinetic energy of a water mass, m, is a measure of its capacity to do work by virtue of its velocity, V. The magnitude of kinetic energy is commonly expressed as $\frac{1}{2}mV^2$. For a pipe flow with mean velocity V and pipe cross-sectional area A, the total mass of water that passes through the cross section in unit time is

$$m = \rho V A$$

The total kinetic energy is

$$\text{K.E.} = \frac{1}{2}mV^2 = \frac{1}{2}(\rho V A)V^2 = \frac{1}{2}\rho A V^3$$

In hydraulics it is the custom to express energy in the units of *energy contained in unit weight of water*. The water mass involved has the weight $W = m \cdot g = \rho A V g$, where g is the gravitational acceleration. The kinetic energy in unit weight of water therefore is

$$\frac{\text{K.E.}}{W} = \frac{\frac{1}{2}\rho A V^3}{\rho g A V} = \frac{V^2}{2g} \tag{3.4}$$

This quantity carries the dimension of a length representing the height of a water column commonly known as the *velocity head*.

The pressure energy is a measure of work done by the pressure force on the water mass. If the pressure at a cross section is P, the pressure force on the entire cross-sectional area must be PA. In a unit time the water mass advances to the right an average distance V, which is the mean velocity, as shown in Figure 3.4. The work done by the pressure force in a unit time is thus $(PA \cdot V)$, which is equal to the total pressure energy:

$$\text{P.E.} = PA \cdot V$$

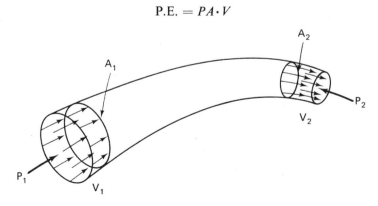

Figure 3.4. Stream tube.

The pressure energy in unit weight of water is

$$\frac{\text{P.E.}}{W} = \frac{PAV}{\rho g AV} = \frac{P}{\rho g} = \frac{P}{\gamma} \qquad (3.5)$$

This quantity also carries the dimension of length; as the velocity head it may be represented by the height of water column known as the *pressure head*.

The elevation energy contained in a water mass is measured by virtue of its elevation above certain reference datum. It is the work required to raise the water thereto. If the water mass is raised h above the datum elevation, the total elevation energy required is

$$\text{E.E.} = W \cdot h$$

The elevation energy contained in the unit weight of water is

$$\frac{\text{E.E.}}{W} = \frac{W \cdot h}{W} = h \qquad (3.6)$$

This is the distance measured from the reference datum to the centerline of the water pipe. The quantity is known as the *elevation head* or *potential head*.

The algebraic sum of the kinetic head, the pressure head, and the elevation head accounts for nearly all the energy contained in unit weight of water flowing through a particular section of a pipe.

Figure 3.5 shows, schematically, the energy heads at two sections along a pipeline. At section 1, the upstream section, the three energy heads are

Figure 3.5. Energy head and head loss in pipe flow.

respectively $V_1^2/2g, P_1/\gamma$, and h_1. The algebraic sum of these three heads gives the point a above the energy datum. The distance measured between points a and b on the energy datum represents the total energy contained in each unit weight of water that passes through section 1.

$$H_1 = \frac{V_1^2}{2g} + \frac{P_1}{\gamma} + h_1 \tag{3.7}$$

During the journey between the two sections, a certain amount of the energy is lost. The remaining total energy in each unit weight of water at section 2 can be measured by the distance between points a' and b' on the energy datum. This distance, as shown in Figure 3.5, is the sum of the kinetic head, the pressure head, and the potential head at section 2.

$$H_2 = \frac{V_2^2}{2g} + \frac{P_2}{\gamma} + h_2 \tag{3.8}$$

The difference in elevations between points a and a' represents the head loss, h_L, between sections 1 and 2. The energy relationship between the two sections can be written in the following form:

$$\frac{V_1^2}{2g} + \frac{P_1}{\gamma} + h_1 = \frac{V_2^2}{2g} + \frac{P_2}{\gamma} + h_2 + h_L \tag{3.9}$$

This relationship is commonly referred to as the *Bernoulli's equation* by hydraulic engineers.

Water flow in pipelines is normally considered incompressible. Thus, in any reach of the pipeline, as shown in Figure 3.5, the water quantity that flows into the reach must equal that flowing out. This condition may be written as

$$Q = AV = A_1 V_1 = A_2 V_2 \tag{3.10}$$

This relationship is commonly known as the *continuity equation*.

Example 3.1

A 25-cm circular pipe carries 0.16 m³/s of water under the pressure of 2000 dyn/cm². The pipe is laid at an elevation of 10.71 m above the mean sea level (MSL) of Freeport, Texas. What is the energy head measured with respect to the MSL?

Solution

The continuity condition, Equation (3.1), requires that

$$Q = AV$$

Hence,

$$V = \frac{Q}{A} = \frac{0.16}{\frac{\pi}{4} \cdot \left(\frac{25}{100}\right)^2} = 3.26 \text{ m/s}$$

The total energy head measured with respect to the MSL is

$$\frac{V^2}{2g} + \frac{P}{\gamma} + h = \frac{(3.26)^2}{2(9.81)} + \frac{2000\left(\frac{10^4}{10^5}\right)}{9810} + 10.71$$

$$= 11.27 \text{ m}$$

Energy loss through friction in the length of a pipeline is commonly termed as the *major loss*, h_f. This is the loss of head due to pipe friction and to the viscous dissipation in flowing water. Several studies have been performed during the past century on laws that govern the loss of head by pipe friction. It has been learned from these studies that resistance to flow in a pipe is:

1. Independent of pressure under which the water flows.
2. Linearly proportional to the pipe length, L.
3. Inversely proportional to some power of the pipe diameter, D.
4. Proportional to some power of the mean velocity, V.
5. Related to the roughness of the pipe, if the flow is turbulent.

Several experimental formulas have been developed in the past. Some of these formulas have been used faithfully in various hydraulic engineering practices.

The most popular pipe flow formula was derived by Henri Darcy (1803–1858), Julius Weisbach (1806–1871), and others about the middle of the nineteenth century. The formula takes the following form:

$$h_f = f\left(\frac{L}{D}\right)\frac{V^2}{2g} \tag{3.11}$$

This formula is commonly known as the Darcy–Weisbach formula. It is conveniently expressed in terms of the velocity head in the pipe. Moreover, it is dimensionally uniform since in engineering practice the *friction factor*, f, is treated as a dimensionless numerical factor; h_f and $V^2/2g$ are both in units of length.

In laminar flow the friction factor f can be determined by balancing the viscous force and the pressure force at the two end sections of a horizontal pipe separated by a distance L. In a cylindrical portion of radius r (see Figure 3.6) the difference in pressure force between the two ends of the cylinder is $(P_1 - P_2) \cdot \pi r^2$ and the viscous force on the cylinder is equal to $2\pi r L \cdot \tau$. The values of shear stress τ have been previously shown, Equation (1.2), to be $\mu(dv/dr)$. Under equilibrium condition, the pressure force and

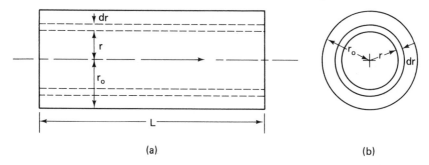

(a)   (b)

Figure 3.6. Geometry of a circular pipe.

the viscous force on the cylinder of water are balanced, we may write

$$-2\pi r L \cdot \left(\mu \frac{dv}{dr} \right) = (P_1 - P_2)\pi r^2$$

The minus sign is used because the velocity decreases as the radial position, r, increases; dv/dr is always negative in a pipe flow. This equation can be integrated to give the general expression of flow velocity in terms of radial position r.

$$v = \frac{P_1 - P_2}{4\mu L}(r_0^2 - r^2) \tag{3.12}$$

where r_0 is the inner radius of the pipe, and the equation shows that the velocity distribution in a laminar pipe flow is a parabolic function of radius, r. The total discharge through the pipe can be obtained by integrating the discharge through the elemental area, $2\pi r \cdot dr$:

$$
\begin{aligned}
Q = \int dQ = \int v \cdot dA &= \int_{r=0}^{r=r_0} \frac{P_1 - P_2}{4\mu L}(r_0^2 - r^2) \cdot 2\pi r \cdot dr \\
&= \frac{\pi r_0^4 (P_1 - P_2)}{8\mu L} = \frac{\pi D^4 (P_1 - P_2)}{128\mu L}
\end{aligned}
\tag{3.13}
$$

This relationship is also known as the Hagen-Poiseuille law* of laminar flow.

*Experimentally derived by G.W. Hagen (1839) and later independently obtained by J.L.M. Poiseuille (1840).

The mean velocity is

$$V = \frac{Q}{A} = \frac{\pi D^4 (P_1 - P_2)}{128 \mu L} \left(\frac{1}{\frac{1}{4} \pi D^2} \right)$$

$$V = \frac{(P_1 - P_2) D^2}{32 \mu L} \tag{3.14}$$

For a horizontal uniform pipe,

$$h_f = \frac{P_1 - P_2}{\gamma}$$

the Darcy–Wiesbach formula may be written as

$$\frac{P_1 - P_2}{\gamma} = f \left(\frac{L}{D} \right) \frac{V^2}{2g} \tag{3.15}$$

Comparing Eqs. (3.14) and (3.15), we have

$$f = \frac{64}{\gamma} \frac{\mu g}{VD}$$

Since $\gamma = \rho g$,

$$f = \frac{64 \mu}{\rho VD} = \frac{64}{N_R} \tag{3.16}$$

indicating a linear relationship between the friction factor, f, and the Reynolds number, N_R, for laminar pipe flow. It is independent of surface roughness of the pipe.

When the Reynolds number approaches a higher value, say, $N_R \gg 2000$, the flow in the pipe becomes practically turbulent and the value of f may then be dependent not only on the Reynolds number but also on the *relative roughness height, e/D*, of the pipe, where e is the measure of the average roughness height of the pipe wall, as previously discussed, and D is the pipe diameter. The roughness height of commercial pipes is commonly described by providing a value of e to the pipe. It means that the pipe has the same value of f at high Reynolds numbers that would be obtained if the pipe were coated with sand grains of a uniform size e. The relative roughness, defined as e/D, for common commercial pipe materials is plotted against the pipe diameter D in Figure 3.7.

It has been determined that immediately next to the pipe wall there exists, a very thin layer of flow commonly referred to as the *laminar sublayer* even though the pipe flow is turbulent. The thickness of the laminar sublayer δ' decreases with an increase in the pipe's Reynolds number. A pipe is said to be *hydraulically smooth* if the average roughness height e is less than the

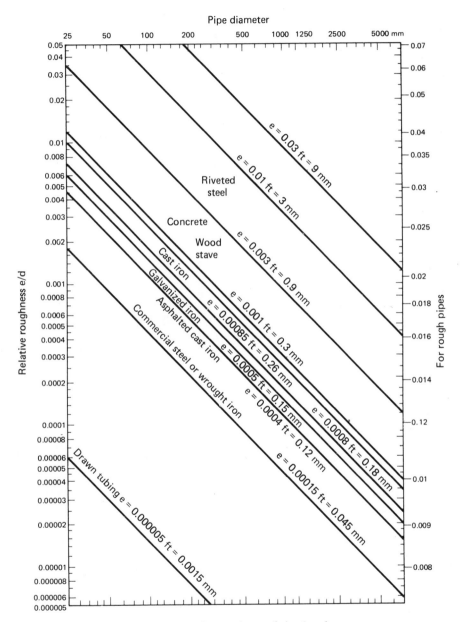

Figure 3.7. Relative roughness of circular pipes.

thickness of the laminar sublayer. In a hydraulically smooth pipe flow the friction factor f is not affected by the surface roughness of the pipe.

Based on laboratory experimental data, it has been found that if $\delta' > 1.7e$, the effect of surface roughness is completely submerged by the laminar sublayer and the pipe flow is hydraulically smooth. In this case, Th. von Karman* developed an equation for the friction factor f:

$$\frac{1}{\sqrt{f}} = 2 \log \left(\frac{N_R \sqrt{f}}{2.51}\right) \tag{3.17}$$

At high Reynolds numbers δ' becomes very small. If $\delta' < 0.08e$, it has been found that the friction factor, f, becomes independent of the Reynolds number and depends only on the relative roughness height. In this case, the pipe behaves as a *hydraulically rough pipe*, and von Karman found that the friction factor f can be expressed as

$$\frac{1}{\sqrt{f}} = 2 \log \left(3.7 \frac{D}{e}\right) \tag{3.18}$$

In between these two extreme cases, if $0.08e < \delta' < 1.7e$, the pipe behaves neither smoothly nor completely roughly. C.F. Colebrook† devised an approximate relationship for this intermediate range:

$$\frac{1}{\sqrt{f}} = -2 \log \left(\frac{\frac{e}{D}}{3.7} + \frac{2.51}{N_R \sqrt{f}}\right) \tag{3.19}$$

Obviously, it is cumbersome to use any of these three equations in engineering practice. A convenient chart was prepared by Lewis F. Moody†† (see Figure 3.8) and is commonly called the *Moody diagram of friction factors for pipe flow.*

The chart clearly shows the four zones of pipe flow: (1) a laminar flow zone where the friction factor f is a simple linear function of the Reynolds number; (2) a critical zone (shaded) where values are uncertain because the flow might be neither laminar nor truly turbulent; (3) a transitional zone where f is a function of both the Reynolds number and the relative roughness height of the pipe wall; and (4) a zone of fully developed turbulence where

*Th. von Karman, "Mechanische Ähnlichkeit und Turbulenz" (Mechanical similitude and turbulence), Proc. 3rd International Congress for Applied Mechanics, Stockholm, Vol. I (1930).
†C.F. Colebrook, "Turbulent flow in pipes, with particular reference to the transition region between smooth and rough pipe laws," *Jour. Inst. Civil Engrs*, London (Feb. 1939).
††L.F. Moody, "Friction factors for pipe flow", Trans. ASME, Vol. 66 (1944).

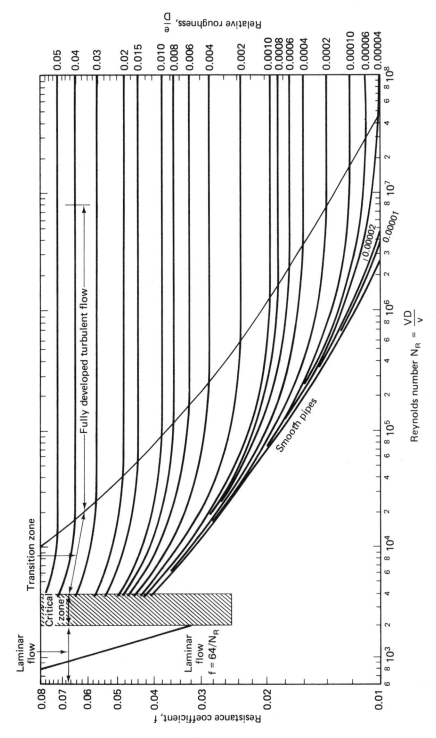

Figure 3.8. Moody chart of resistance for uniform flow in circular pipes.

the value of f depends solely on the relative roughness height and is independent of the Reynolds number. Figure 3.8 may be used together with Figure 3.7 to obtain the friction factor f for circular pipes.

Example 3.2

Compute the discharge capacity of a 3-m wood-stave pipe in its best condition carrying water at 10°C. It is allowed to have a head loss of 2 m/km of pipe length.

Solution

From Eq. (3.11), the friction head loss in the pipe is,

$$h_f = f\left(\frac{L}{D}\right)\frac{V^2}{2g}$$

Hence,

$$2 = f\left(\frac{1000}{3}\right)\frac{V^2}{2(9.81)}$$

$$fV^2 = \frac{2 \cdot 3 \cdot 2 \cdot 9.81}{1000} = 0.12 \qquad (1)$$

From Figure 3.7 we obtain for the 3-m wood-stave pipe:

$$\frac{e}{D} = 0.0001$$

At 10°C the kinematic viscosity of water is $v = 1.31 \cdot 10^{-6}$ m²/s. Therefore,

$$N_R = \frac{DV}{v} = \frac{3V}{1.31 \cdot 10^{-6}} = 2.29 \cdot 10^6 V \qquad (2)$$

By using Figure 3.8, Eqs. (1) and (2) are solved by iteration until both conditions are satisfied. The iteration procedure is demonstrated as follows:

Assume $f = 0.02$. From Eq. (1) we get $V = 2.45$ m/s, and from Eq. (2) we get $N_R = 5.6 \times 10^6$. This number is taken to the Moody Chart (Figure 3.8) to obtain the value $f = 0.0122$, which is different from the assumed value. For the second trial, let $f = 0.0122$; we get $V = 3.14$ m/s and $N_R = 7.2 \times 10^6$. The Moody Chart gives $f = 0.0121$ which is considered close enough to the assumed value and will be used to calculate the flow velocity,

$$V = 3.15 \text{ m/s}$$

This value and the cross-sectional area gives the discharge.

$$Q = AV = \pi\left(\frac{3^2}{4}\right)(3.15)$$

$$= 22.27 \text{ m}^3/\text{s}$$

Example 3.3

Estimate the size of a uniform, horizontal welded-steel pipe installed to carry 500 l/s of water at 10°C. The allowable pressure loss is 5 m/km of pipe length.

Solution

From Eq. (3.11)

$$h_f = f\frac{L}{D}\frac{V^2}{2g}$$

where $L = 1000$ m, $h_f = 5$ m, and

$$V = \frac{Q}{A} = \frac{4Q}{\pi D^2} = \frac{4\left(500 \cdot \frac{1}{1000}\right)}{\pi D^2} = \frac{0.64}{D^2}$$

Therefore,

$$h_f = f\frac{1000}{D}\frac{\left(\frac{0.64}{D^2}\right)^2}{2(9.81)}$$

$$5 = 20.88\frac{f}{D^5} \tag{1}$$

At 10°C, $v = 1.310 \cdot 10^{-6}$ m²/s.

$$N_R = \frac{DV}{v} = \frac{D\left(\frac{0.64}{D^2}\right)}{1.310 \cdot 10^{-6}} = \frac{4.88 \cdot 10^5}{D} \tag{2}$$

Equations (1) and (2) may be solved by trial by using both Figure 3.7 and Figure 3.8 as follows: Let $D = 0.8$ m. Eq. (1) gives $f = 0.078$; Eq. (2) gives $N_R = 6.1 \times 10^5$; and Figure 3.7 gives $e/D = 0.0015$. Applying these values to Figure 3.8, we get $f = 0.0125$. This value does not agree with the value obtained from Eq. (1). We shall try another diameter. Let $D = 0.6$ m, $f = 0.019$, $N_R = 8.1 \times 10^5$, and $e/D = 0.002$, from which we obtain $f = 0.015$. The two values are closer now, but they still are not close enough for an accurate solution. We shall try again. Let $D = 0.57$ m, $f = 0.014$, $N_R = 8.56 \times 10^5$, $e/D = 0.002$, from which we have $f = 0.014$. The diameter of the pipe is 0.57 m.

PROBLEMS

3.2.1. A 30-cm circular cast-iron pipeline, 2 km long, carries water at 10°C. What is the maximum discharge if a 4.6-m head loss is allowed?

3.2.2. A new asphalted cast-iron pipe, 45 cm in diameter, carries water from a reservoir to a location 1.37 km away and discharges the water into the air. If the entrance to the pipe is 3 m below the water surface in the reservoir and the pipe is laid on a 1/100 slope, determine the discharge at 10°C.

3.2.3. A horizontal commercial steel pipe, 0.5 m in diameter, carries 0.4 m³/s of water at 20°C. Calculate the pressure change in the pipe per kilometer length.

3.2.4. A 6-cm galvanized iron pipe is installed on a 1/50 slope and carries water at 85°C. What is the pressure drop in the 20-m-long pipe when the discharge is 8.2 l/s?

3.2.5. Two sections, *A* and *B*, are 5 km apart along a 4-m riveted-steel pipe in its best condition. *A* is 100 m higher than *B*. If the water temperature is 15°C and the pressure heads measured at *A* and *B* are 8.3 m and 76.7 m, respectively, what is the flow rate?

3.2.6. The city water company wants to transport 1800 m³ of water per day to a plant from a reservoir 8 km away. The water surface elevation at the reservoir is 5 m above the entrance of the pipe, and the pipe may be laid on a 1/500 slope. What is the minimum diameter of a concrete pipe that may be used if the water temperature varies between 5°C and 20°C? Assume pipe discharges into open air and neglect minor losses.

3.3   EMPIRICAL FORMULAS FOR WATER FLOW IN PIPES

Throughout the history of civilization hydraulic engineers have designed and built systems to deliver water for human usage. Many of these designs, in the earlier days, were based on empirical formulas. Generally speaking, empirical design formulas were developed from the experiences of dealing with fluid flow under certain conditions in a specific range. They do not, normally, have a sound analytical basis. For this reason, the empirical formulas may not be dimensionally correct, and, if so, can only be applicable to the conditions and ranges specified.

One of the best examples is the Hazen–Williams formula, which was developed for water flow in larger pipes ($D \geq 5$ cm) within a moderate range of water velocity ($V \leq 3$ m/s). This formula has been used extensively for the designing of water-supply systems in the United States. The Hazen–Williams formula, originally developed for the British measurement system, has been written in the form

$$V = 1.318 C_{HW} R_H^{0.63} S^{0.54} \tag{3.20}$$

where S is the slope of the *energy gradient line* (EGL), or head loss per unit length of the pipe ($S = h_f/L$), and R_H is the *hydraulic radius*, defined as the water cross-sectional area, A, divided by the *wetted perimeter*, P. For a circular pipe, $A = \pi D^2/4$ and $P = \pi D$, the hydraulic radius is

$$R_H = \frac{A}{P} = \frac{\dfrac{\pi D^2}{4}}{\pi D} = \frac{D}{4} \tag{3.21}$$

The Hazen–Williams coefficient, C_{HW}, is not a function of the flow conditions (i.e., Reynolds number). Its values range from 140 for very smooth, straight pipe down to 90 or 80 for old, unlined tuberculated pipe. Generally, the value of 100 is taken for average conditions. The values of C_{HW} for commonly used water-carrying conduits is listed in Table 3.1.

TABLE 3.1 Hazen–Williams Coefficient, C_{HW}, for Different Types of Pipe

Pipe Materials	C_{HW}
Asbestos Cement	140
Brass	130-140
Brick sewer	100
Cast-iron	
New, unlined	130
10 yr. old	107-113
20 yr. old	89-100
30 yr. old	75- 90
40 yr. old	64- 83
Concrete or concrete lined	
Steel forms	140
Wooden forms	120
Centrifugally spun	135
Copper	130-140
Galvanized iron	120
Glass	140
Lead	130-140
Plastic	140-150
Steel	
Coal-tar enamel lined	145-150
New unlined	140-150
Riveted	110
Tin	130
Vitrified clay (good condition)	110-140
Wood stave (average condition)	120

Note that the Hazen–Williams formula as shown in Eq. (3.20) is applicable *only* for the British units in which the velocity is measured in feet per second and the hydraulic radius R is measured in feet. When used in S. I. units, the Hazen–Williams formula may be written in the following form:

$$V = 0.85 C_{HW} R_H^{0.63} S^{0.54} \qquad (3.22)$$

Solution for the Hazen–Williams formula, Eq. (3.22), may be obtained from the chart provided in Figure 3.9. The following example demonstrates the use of the chart.

Figure 3.9. Solution chart for the Hazen–Williams formula, $V = 0.85 C_{HW} R_H^{0.63} S^{0.54}$.

Example 3.4

A 100-m-long pipe with $D = 20$ cm and $C_{HW} = 100$ carries a discharge of 30 l/s. Determine the head loss in the pipe.

Solution

Applying the conditions given to the solution chart, Figure 3.9, the energy slope is obtained:

$$S = 0.0058 \text{ m/m}$$

The energy slope is defined as the energy loss per unit pipe length, or

$$S = \frac{h_f}{L} = \frac{h_f}{100} = 0.0058 \text{ m/m}$$

Hence,

$$h_f = 0.0058 \text{ m/m} \cdot 100 \text{ m} = 0.58 \text{ m}$$

Another popular empirical formula is the *Manning formula* developed and used extensively for open channel designs (discussed in detail in Chapter 5). Also quite commonly used for pipe flows, the Manning formula may be expressed in the following form:

$$V = \frac{1}{n} R_H^{2/3} S^{1/2} \tag{3.23}$$

where n is Manning's coefficient of roughness, specifically known to hydraulic engineers as Manning's n.

In English units, the Manning formula is written as

$$V = \frac{1.49}{n} R_H^{2/3} S^{1/2} \tag{3.24}$$

where the hydraulic radius, R_H, is measured in feet, and the velocity is measured in units of feet per second. Table 3.2 contains typical values of n for water flow in common pipes.

TABLE 3.2 Manning's Roughness Coefficient, *n*, for Pipe Flows

	Manning's n	
Type of Pipe	*Min.*	*Max.*
Glass, brass, or copper	0.009	0.013
Smooth cement surface	0.010	0.013
Wood-stave	0.010	0.013
Vitrified sewer pipe	0.010	0.017
Cast-iron	0.011	0.015
Concrete, precast	0.011	0.015
Cement mortar surfaces	0.011	0.015
Common-clay drainage tile	0.011	0.017
Wrought iron	0.012	0.017
Brick with cement mortar	0.012	0.017
Riveted-steel	0.017	0.020
Cement rubble surfaces	0.017	0.030
Corrugated metal storm drain	0.020	0.024

Solution to the Manning formula may also be obtained from the chart provided in Figure 3.10. The following example demonstrates the use of Figure 3.10 for solution of the Manning formula.

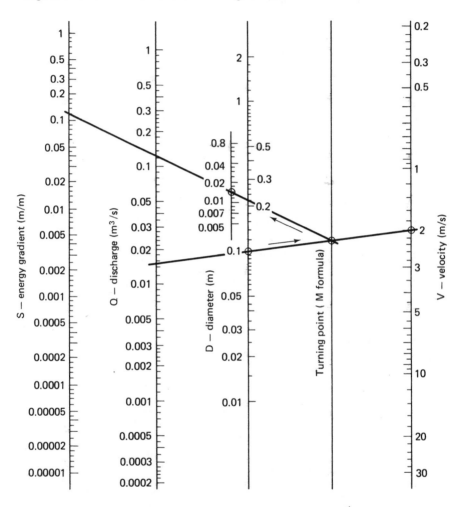

Figure 3.10. Solution chart for the Manning formula, $V = \dfrac{1}{n} R_H^{2/3} S^{1/2}$.

Example 3.5

A horizontal pipe with 10 cm uniform diameter is 200 m long. Manning's roughness coefficient is $n = 0.015$ and the measured pressure drop is 24.6 m in water column. Determine the discharge.

Solution

The energy gradient in the pipe is

$$S = \frac{h_f}{L} = \frac{24.6 \text{ m}}{200 \text{ m}} = 0.123$$

Applying this value and the conditions given above to the solution chart in Figure 3.10 we obtain the following discharge:

$$Q = 0.0157 \text{ m}^3/\text{s} = 15.7 \text{ l/s}$$

PROBLEMS

3.3.1. A 6-km-long, new cast-iron pipeline carries 320 *l*/s of water at 30°C. The pipe diameter is 30 cm. Compare the head loss calculated from (a) the Hazen–Williams formula, (b) the Manning formula, and (c) the Darcy–Weisbach formula.

3.3.2. An 80-cm riveted-steel pipe carries water at 10°C from reservoir A to reservoir B. The length of the pipe is 3.2 km. The elevation difference between the two reservoirs is 102 m. Calculate the discharge by (a) the Hazen–Williams formula, (b) the Manning formula, and (c) the Darcy–Weisbach formula.

3.3.3. Repeat the calculation in Prob. 3.2.4 but use (a) the Manning formula, and (b) the Hazen–Williams formula, and compare the results with those obtained from the Darcy-Weisbach formula. Discuss the differences.

3.3.4. Two reservoirs 1200 m apart are connected by a 50-cm smooth concrete pipe. If the two reservoirs have an elevation difference of 5 m, determine the discharge in the pipe by (a) the Hazen–Williams formula, (b) the Manning formula, and (c) the Darcy–Weisbach formula.

3.3.5. Research in the library and find, in addition to the three introduced in this section, six empirical formulas involving head loss in pipe lines. List the author(s) and limitations of each formula.

3.4   LOSS OF HEAD DUE TO CONTRACTION

Sudden contraction in a pipe usually causes a marked drop in pressure in the pipe due to both the increase in velocity and the loss of energy to turbulence. The phenomenon of a sudden contraction is schematically represented in Figure 3.11.

The vertical distance measured between the energy gradient line (EGL) and the pipe centerline represents the total energy head at any particular location along the pipe. The vertical distance measured between the hydraulic gradient line (HGL) and the pipe centerline represents the pressure head,

Figure 3.11. Head loss and pressure variation due to sudden contraction.

P/γ, and the distance between the EGL and HGL is the velocity head, $V^2/2g$, at the location. After point B the HGL begins to drop as the stream picks up speed and a region of stagnant water appears at the corner of contraction C. Immediately downstream from the contraction the streamlines separate from the pipe wall and form a high-speed jet that reattaches to the wall at point D. The phenomenon that takes place between C and D is known to hydraulic engineers as the *vena contracta*, which will be discussed in detail in Chapter 9. Most of the energy loss in a pipe contraction takes place between C and D where the jet stream velocity is high and the pressure is low. A certain amount of pressure recovers between D and E as the jet stream gradually dissipates and normal pipe flow resumes. Downstream from point E the EGL and HGL lines become again parallel to each other, but they take a steeper slope where a higher rate of energy dissipation in the smaller pipe is expected.

The loss of head in a sudden contraction may be represented in terms of velocity head in the downstream pipe as

$$h_c = K_c \left(\frac{V^2}{2g}\right) \tag{3.25}$$

Here, K_c is the coefficient of contraction. Its value varies with the ratio of contraction, D_2/D_1, as shown in Table 3.3.

Head loss due to pipe contraction may be greatly reduced by introducing a gradual pipe transition known as a *confusor* and shown in Figure 3.12. The head loss in this case may be expressed as

$$h_c = \frac{K_c'}{2g}(V_2^2 - V_1^2) \tag{3.26}$$

TABLE 3.3 Values of the Coefficient K_c for Sudden Contraction

Velocity in Smaller Pipe V(m/s)	Ratio of Smaller to Larger Pipe Diameters, D_2/D_1									
	0.0	0.1	0.2	0.3	0.4	0.5	0.6	0.7	0.8	0.9
1	0.49	0.49	0.48	0.45	0.42	0.38	0.28	0.18	0.07	0.03
2	0.48	0.48	0.47	0.44	0.41	0.37	0.28	0.18	0.09	0.04
3	0.47	0.46	0.45	0.43	0.40	0.36	0.28	0.18	0.10	0.04
6	0.44	0.43	0.42	0.40	0.37	0.33	0.27	0.19	0.11	0.05
12	0.38	0.36	0.35	0.33	0.31	0.29	0.25	0.20	0.13	0.06

Figure 3.12. Pipe confusor.

The values of K'_c vary with the transition angle, α:

α	10°	20°	30°	40°
k'_c	0.20	0.28	0.32	0.35

The loss of head at the entrance to a pipe from a large reservoir is a special case of loss of head due to contraction. Since the water cross-sectional area in the reservoir is very large compared to that of the pipe, a ratio of contraction of zero may be taken. For a square-cornered entrance, where the entrance of the pipe is flush with the reservoir wall, as shown in Figure 3.13(a), the K_c values shown for $D_2/D_1 = 0.0$ in Table 3.3 are used.

The general formula for head loss at the entrance to a pipe is also expressed in terms of the velocity head of the pipe.

$$h_e = K_e \left(\frac{V^2}{2g}\right) \qquad (3.27)$$

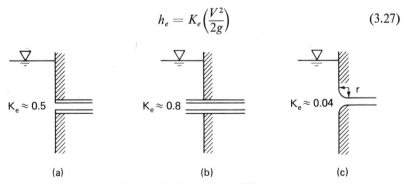

Figure 3.13. Entrance conditions.

The approximate values for the coefficient of entrance loss K_e for different entrance conditions are shown in Figure 3.13(a), (b), and (c).

3.5   LOSS OF HEAD DUE TO ENLARGEMENT

The behavior of the energy gradient line and the hydraulic gradient line in the vicinity of a sudden pipe enlargement is schematically demonstrated in Figure 3.14. At the corner of the sudden enlargement A the streamline separates from the wall of the large pipe and leaves an area of relative stagnancy between A and B, in which a large recirculating eddy is formed to fill the space. Most of the energy loss in a sudden enlargement takes place between A and B where the streamlines reattach to the wall. A recovery of pressure may take place here as a result of the decrease of velocity in the pipe. The high-speed jet stream gradually slows down and reaches an equilibrium at point C. From this point downstream the normal pipe flow conditions resume and the energy gradient line takes on a smaller slope than that of the approaching pipe, as expected.

The loss of head due to sudden enlargement in a pipe can be derived from the momentum considerations (e.g., Daugherty and Franzini, 1976*). The magnitude of the head loss may be expressed as

$$h_e = \frac{V_1^2 - V_2^2}{2g} \tag{3.28}$$

It means that, physically, the change in velocity heads between the two sections of pipe is the head loss in the sudden expansion.

Figure 3.14. Head loss due to sudden enlargement.

*Daugherty, R.L. and J.B. Franzini, *Fluid Mechanics with Engineering Applications*, McGraw-Hill Book Company, New York, (1977).

The head loss due to pipe enlargement may be greatly reduced by introducing a gradual pipe transition known as a *diffusor* and shown in Figure 3.15. The head loss in this pipe transition case may be expressed as

$$h_e = K_e \frac{V_1^2 - V_2^2}{2g} \tag{3.29}$$

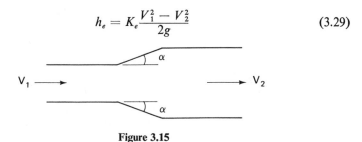

Figure 3.15

The values of K_e vary with the diffusor angle α:

α	10°	20°	30°	40°
K_e	0.39	0.80	1.00	1.06

A submerged pipe discharging into a large reservoir is a special case of loss of head due to enlargement. The flow velocity V in the pipe is discharged from the end of a pipe into a reservoir that is so large that the velocity within it is almost zero. From Eq. (3.28) we see that the entire velocity head of the pipe flow is dissipated and that the *discharge head loss* is

$$h_e = \frac{V^2}{2g} \tag{3.30}$$

The phenomenon of discharge loss is shown in Figure 3.16.

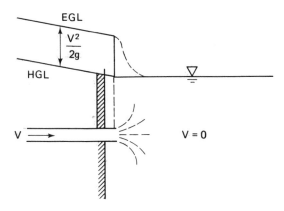

Figure 3.16. Head loss of discharge.

3.6 LOSS OF HEAD IN BENDS

Pipe flow around a bend experiences an increase of pressure along the outer wall and a decrease of pressure along the inner wall. A certain distance downstream from the bend the velocity and pressure resume normal distributions. In order to achieve this, the inner wall pressure must rise to bring the velocity back to the normal value. The velocity near the inner wall of the pipe is lower than that at the outer wall and it must also increase to the normal value. The simultaneous demand of energy may cause separation of the stream from the inner wall, as shown in Figure 3.17(a). In addition, the unbalanced pressure at the bend causes secondary current, as shown in Figure 3.17(b). This transverse current and the axial velocity form a pair of spiral flows that persist as far as 100 diameters downstream from the bend. Thus, the head loss at the bend is combined with the distorted flow conditions downstream from the bend until the spiral flows are dissipated by viscous friction.

The head loss produced at a bend was found to be dependent on the ratio of the radius of curvature of the bend, R, to the diameter of the pipe, D. Since the spiral flow produced by a bend extends some distance downstream from the bend, the head loss produced by different pipe bends placed close together cannot be treated by simply adding the losses of each one separately. The total loss of a series of bends placed close together depends not only on the spacing between the bends but also on the direction of the bends. Detailed analysis of head loss produced by a series of bends is a rather complex matter for which much research has yet to be done.

In hydraulic design the loss of head due to a bend, in excess of that which would occur in a straight pipe of equal length, may be expressed in terms of the velocity head as

$$h_b = K_b \frac{V^2}{2g} \qquad (3.31)$$

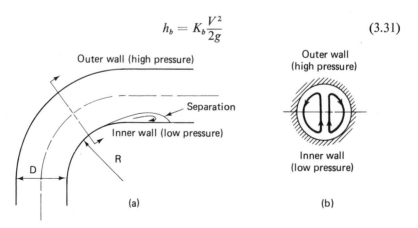

Figure 3.17. (a) Separation at a bend. (b) Secondary flow at a bend.

For a smooth pipe bend of 90°, the values of K_b for various values of R/D as determined by Beij* are listed in Table 3.4. The bend loss has also been

TABLE 3.4

R/D	1	2	4	6	10	16	20
K_b	0.35	0.19	0.17	0.22	0.32	0.38	0.42

found nearly proportional to the angle of bend, α, for pipe bends other than 90° in steel pipes and drawn tubings.

3.7 LOSS OF HEAD THROUGH VALVES

Valves are installed in pipelines to control flow by imposing high head losses through the valves. Depending on how the particular valve is designed, a certain amount of energy loss usually takes place even when the valve is fully open. Similar to all other losses in pipes, the head loss through valves may also be expressed in terms of velocity head in the pipe:

$$h_v = K_v \frac{V^2}{2g} \tag{3.32}$$

The values of K_v vary with the design of the valves. From the viewpoint of designing a hydraulic system, it is necessary to determine the required loss of head through a valve in the system. The values of K_v for common valves are listed in Table 3.5.

Example 3.6

Figure 3.18 shows two sections of cast-iron pipe connected in series bring water from a reservoir and discharge it into air at a location 100 m below the water surface elevation in the reservoir through a rotary valve. If the water temperature is 10°C, and square connections are used, determine the discharge.

Solution

The total available energy, 100 m of water column, is equal to the velocity head at the discharge end plus all the head losses incurred in the pipeline system. These relationships, as shown in Figure 3.18, may be expressed as follows:

$$h_e + h_{f_1} + h_c + h_{f_2} + h_v + \frac{V_2^2}{2g} = 100$$

*K. H. Beij, "Pressure Losses for Fluid Flow in 90° Pipe Bends," *Jour. Research Natl. Bur. Standards*, 21 (1938).

TABLE 3.5 Values of K$_v$ for Certain Common Hydraulic Valves

A. Gate valves

Closed

Open

K$_V$ = 0.15 (fully open)

B. Globe valves:

Closed

Open

C. Check valves:

Closed

Hinge (Swing type)

Open

Swing type K$_V$ = 2.5   (fully open)
Ball type   K$_V$ = 70.0 (fully open)
Lift type   K$_V$ = 12.0 (fully open)

D. Rotary valves:

Closed

Open

K$_V$ = 10.0 (fully open)

Figure 3.18

where h_e is the head loss at entrance. For square-cornered entrance, Eq. (3.27) gives:

$$h_e = (0.5)\frac{V_1^2}{2g}$$

h_{f_1} is the head loss due to friction in pipe section 1-2. From Eq. (3.11),

$$h_{f_1} = f_1\frac{1000}{0.40}\frac{V_1^2}{2g}$$

h_c is the head loss due to sudden contraction at section 2. From Table 3.3 and Eq. (3.25),

$$h_c = K_c\frac{V_2^2}{2g} = 0.33\frac{V_2^2}{2g}$$

h_{f_2} is the head loss due to friction in pipe section 2-3:

$$h_{f_2} = f_2\frac{1200}{0.20}\frac{V_2^2}{2g}$$

h_v is the head loss at the valve. From Table 3.5 and Eq. (3.32)

$$h_v = K_v\frac{V_2^2}{2g} = (10)\frac{V_2^2}{2g}$$

Therefore,

$$100 = \left(1 + 10 + f_2\frac{1200}{0.20} + 0.33\right)\frac{V_2^2}{2g} + \left(f_1\frac{1000}{0.40} + 0.5\right)\frac{V_1^2}{2g}$$

From the continuity equation (3.10), we have

$$A_1V_1 = A_2V_2$$

$$\frac{\pi}{4}(0.4)^2V_1 = \frac{\pi}{4}(0.2)^2V_2$$

The above relations give

$$V_2^2 = \frac{1962}{11.36 + 156.25 f_1 + 6000 f_2}$$

To evaluate f_1 and f_2, we have:

$$N_{R_1} = \frac{D_1 V_1}{\nu} = \frac{0.4}{1.31 \cdot 10^{-6}} V_1 = 7.63 \cdot 10^4 \cdot V_2$$

$$N_{R_2} = \frac{D_2 V_2}{\nu} = \frac{0.2}{1.31 \cdot 10^{-6}} V_2 = 1.53 \cdot 10^5 \cdot V_2$$

where $\nu = 1.31 \cdot 10^{-6}$ at 10°C. For the 40-cm pipe, $e/D = 0.00065$, which yields $f_1 \approx 0.0178$. For the 20-cm pipe, $e/D = 0.0013$; so $f_2 \approx 0.0205$. Solving the above equation for V_2 provides the following:

$$V_2^2 = \frac{1962}{11.36 + 156.25(0.0178) + 6000(0.0205)}$$

$$V_2 = 3.78 \text{ m/s}$$

$$V_1 = 0.25(3.78) = 0.94 \text{ m/s}$$

Hence, $N_{R_1} = 7.63 \cdot 10^4 (3.78) = 2.88 \cdot 10^5; \qquad f_1 = 0.019$

$N_{R_2} = 1.53 \cdot 10^5 (3.78) = 5.78 \cdot 10^5; \qquad f_2 = 0.021$

These f values do not agree with those stated previously, a second trial must be made assuming that $f_1 = 0.019$ and $f_2 = 0.021$. Repeating the above calculation, we get $V_2 = 3.74$, $N_{R_1} = 2.9 \cdot 10^5$, $N_{R_2} = 5.72 \cdot 10^5$. From Figure 3.7, we have,

$$f_1 \cong 0.019; \qquad f_2 \cong 0.021.$$

Therefore, the discharge is

$$Q = A_2 V_2 = \frac{\pi}{4}(0.2)^2(3.74) = 0.12 \text{ m}^3/\text{s}$$

PROBLEMS

3.7.1. A 75-m-long cast-iron pipe, 15 cm in diameter, takes water from a lake at 2 m below the lake surface. The pipe is laid on a 1/25 slope and has a 90° bend ($R/D = 1.0$) at midlength. What is the flow rate if the pipe discharges freely into the air, if the water temperature in the pipe is 20°C?

3.7.2. A globe valve is installed in a 1-m steel penstock which brings water from a high dam to a hydroelectric power plant. The entrance is 30 m below the reservoir water surface elevation, and the penstock, 440 m long, is laid on a 1/2 slope. Determine the maximum discharge.

3.7.3. A 30-m-high water tower supplies water to a residential area by means of a 20-cm diameter, 800-m-long steel pipe. In order to preserve maximum pressure head at the delivery, it is considered to replace 94% of the pipe length with a larger (30 cm diameter) steel pipe connected by a 30° confusor. If the peak discharge demand is 0.25 m³/s, find out how much pressure head it may gain at temperature 20°C.

3.7.4. A pump moves water from a large sump 3 m deep at a rate of 18 *l*/s. The static water level in the sump is 1.5 m above the intake (5-cm steel pipe) which is located 50 cm above the sump bottom. What suction pressure must the pump provide? The total head loss from the sump to the pump is approximately 1.5 times the velocity head.

3.7.5. A cylindrical tank 5 m in diameter is filled with water to 3 m deep. A short horizontal pipe with a 20-cm diameter and a lift-type check valve is used to drain the tank from the bottom. How long does it take to drain 50% of the tank?

3.8 PIPELINE CONNECTING TWO RESERVOIRS

In previous sections we examined the head losses in pipe flows under various conditions. The interest of engineers, however, is to apply these fundamentals to solve practical problems. Pipe flow problems may generally be classified into four major groups: (1) pipelines, (2) pipe branchings, (3) pipe networks, and (4) siphon and pump in pipelines. Each of these problems will be discussed in the following sections.

A *pipeline* is a connection of one or more pipes in series and is designed to transport water from one reservoir to another. There are three principal types of pipeline problem: (1) Given the flow rate and the pipe combinations, determine the total head loss. (2) Given the allowable total head loss and the combination of the pipes, determine the flow rate. (3) Given the flow rate and the allowable total head loss, determine the pipe diameter. The first type of problem can be solved by the direct approach, but the second and third types involve trial-and-error procedures as shown in the following examples.

Example 3.7

Two new cast-iron pipes in series connect two reservoirs (see Figure 3.19). Both pipes are 300 m long and have diameters of 0.6 m and 0.4 m, respec-

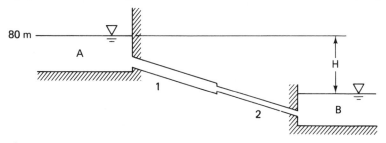

Figure 3.19

tively. The elevation of water surface in reservoir A is 80 m. The discharge of
10°C water from reservoir A to reservoir B is 0.5 m³/s. Find the elevation of
the surface of reservoir B. Assume a sudden contraction at the junction and a
square-edge entrance.

Solution

Designating subscript 1 for the upstream pipe and subscript 2 for the downstream pipe, we may write

$$V_1 = \frac{Q}{A_1} = \frac{0.5}{\frac{1}{4} \cdot \pi \cdot (0.6)^2} = 1.77 \text{ m/s}$$

$$V_2 = \frac{Q}{A_2} = \frac{0.5}{\frac{1}{4} \cdot \pi \cdot (0.4)^2} = 3.98 \text{ m/s}$$

$$N_{R_1} = \frac{V_1 D_1}{v} = \frac{1.77(0.6)}{1.31 \cdot 10^{-6}} = 8.08 \cdot 10^5$$

$$N_{R_2} = \frac{V_2 D_2}{v} = \frac{3.98(0.4)}{1.31 \cdot 10^{-6}} = 1.22 \cdot 10^6$$

From Figure 3.6 we have

$$\frac{e}{D_1} = 0.00043$$

$$\frac{e}{D_2} = 0.00065$$

From Figure 3.7 we have

$$f_1 = 0.017 \quad \text{and} \quad f_2 = 0.018$$

The total energy head $= h_e + h_{f_1} + h_c + h_{f_2} + h_d$. From Eqs. (3.11),
(3.27), (3.25), and (3.30) we may write

$$H = \left(0.5 + f_1 \frac{L_1}{D_1}\right)\frac{V_1^2}{2g} + \left(0.24 + f_2 \frac{L_2}{D_2} + 1\right)\frac{V_2^2}{2g}$$

$$= 13.34 \text{ m}$$

The elevation of the surface of reservoir B is

$$80 - 13.34 = 66.66 \text{ m}$$

Example 3.8

Pipeline AB connects two reservoirs. The difference in elevation between the two reservoirs is 10 m. The pipeline consists of an upstream section, $D_1 = 0.75$ m and $L_1 = 1500$ m, and a downstream section, $D_2 = 0.5$ m and $L_2 = 1000$ m. The pipes are cast-iron and are connected end-to-end with a sudden reduction of area. Assume the water temperature at $10°C$. Compute the discharge capacity.

Solution

Since the discharge is yet unknown, the velocity in each pipe can only be assumed to be V_1 and V_2, respectively. The total energy equation, as above, will contain these two assumed quantities. It cannot be solved directly. A trial procedure is therefore used. For water temperature at $10°C$, $v = 1.31 \cdot 10^{-6}$ m²/s. The corresponding Reynolds numbers may be expressed as:

$$N_{R_1} = \frac{V_1 D_1}{v} = \frac{V_1(0.75)}{1.31 \cdot 10^{-6}} = 5.73 \cdot 10^5 \cdot V_1 \tag{1}$$

$$N_{R_2} = \frac{V_2 D_2}{v} = \frac{V_2(0.5)}{1.31 \cdot 10^{-6}} = 3.82 \cdot 10^5 \cdot V_2 \tag{2}$$

From the continuity condition, $A_1 V_1 = A_2 V_2$, we have

$$\frac{\pi}{4}(0.75)^2 \cdot V_1 = \frac{\pi}{4}(0.5)^2 \cdot V_2$$

$$V_2 = 2.25 V_1 \tag{3}$$

Substituting Eq. (3) into Eq. (2), we get

$$N_{R_2} = (3.82 \cdot 10^5) \cdot (2.25 V_1) = 8.60 \cdot 10^5 \cdot V_1 \tag{4}$$

The energy equation may be written as

$$10 = \left[0.5 + f_1\left(\frac{1500}{0.75}\right)\right]\frac{V_1^2}{2g} + \left[0.24 + f_2\left(\frac{1000}{0.5}\right) + 1\right]\frac{V_2^2}{2g}$$

$$10 = (0.35 + 101.94 f_1 + 516.05 f_2)V_1^2 \tag{5}$$

From Figure 3.6, $e/D_1 = 0.00035$ and $e/D_2 = 0.00052$. As a first trial, we assume $f_1 = 0.015$ and $f_2 = 0.016$.

$$10 = [0.35 + 101.94(0.015) + 516.05(0.016)]V_1^2$$

$$V_1 = 0.99 \text{ m/s}$$

$$N_{R_1} = 5.73 \cdot 10^5 (0.99) = 5.67 \cdot 10^5$$
$$N_{R_2} = 8.60 \cdot 10^5 (0.99) = 8.51 \cdot 10^5$$

From Figure 3.7, $f_1 = 0.165$ and $f_2 = 0.0172$. These values do not agree with the values stated previously. For the second trial, assume that $f_1 = 0.0165$ and $f_2 = 0.0175$.

$$V_1 = 0.95 \text{ m/s}; \qquad N_{R_1} = 5.44 \times 10^5; \qquad N_{R_2} = 8.17 \times 10^5$$

We obtain, from Figure 3.7, $f_1 = 0.0165$ and $f_2 = 0.0175$. These values are the same as the assumed values, suggesting that $V_1 = 0.95$ m/s is the actual velocity in the upstream pipe. Therefore, the discharge is

$$Q = A_1 V_1 = \frac{\pi}{4}(0.75)^2(0.95) = 0.42 \text{ m}^3/\text{s}$$

Example 3.9

A concrete pipeline is installed to deliver 6 m³/s of water between two reservoirs 17 km apart. If the elevation difference between the two reservoirs is 12 m, determine the diameter of the pipe.

Solution

The mean velocity may be obtained by using the continuity condition, Eq. (3.10):

$$V = \frac{Q}{A} = \frac{6}{\frac{\pi}{4}D^2} = \frac{7.64}{D^2}$$

and,

$$N_R = \frac{DV}{\nu} = \frac{D\left(\frac{7.64}{D^2}\right)}{1.31 \cdot 10^{-6}} = \frac{5.83 \cdot 10^6}{D}$$

Neglect the minor losses (for long pipes of $L/D \gg 1000$, minor losses may be neglected). The energy loss contains only the friction loss term. Eq. (3.11) gives

$$12 = f\left(\frac{L}{D}\right)\frac{V^2}{2g} = f\left(\frac{L}{D}\right)\frac{Q^2}{2gA^2} = f\left(\frac{17{,}000}{D}\right) \cdot \left(\frac{6^2}{2 \cdot 9.81 \cdot \left(\frac{\pi}{4}\right)^2 \cdot D^4}\right)$$

Simplified,

$$0.000237 = \frac{f}{D^5} \tag{a}$$

For concrete pipes, $e = 0.3$ mm to $e = 3.0$ mm, use $e = 1.50$ mm, and assume that $D = 2.5$ m for the first trial.

$$\frac{e}{D} = 0.0006$$

and $N_R = 2.33 \cdot 10^6$. From Figure 3.7 we obtain $f = 0.0174$. Substituting these values into Eq. (a), we see that

$$0.00023 \neq \frac{0.0174}{(2.5)^5} = 0.00018$$

Hence, a different pipe diameter must be assumed for the second trial. Assume $D = 2.4$ m. Hence, $e/D = 0.000625$ and $N_R = 2.43 \cdot 10^6$. From Figure 3.7, we obtain $f = 0.0176$. Substituting the above values again into Eq. (a), we now have

$$0.00023 \cong \frac{0.00176}{(2.4)^5} = 0.00022$$

The value on the right-hand side is considered close enough to that on the left-hand side and the pipe diameter of 2.40 m is selected.

PROBLEMS

3.8.1. Sketch the energy gradient line (EGL) and the hydraulic gradient line (HGL) for the compound pipe shown in Figure 3.20. Consider all the losses and change of velocity and pressure heads.

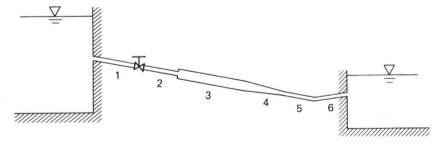

Figure 3.20

3.8.2. A 40-cm straight pipe is used to carry water at 20°C from one reservoir to another reservoir which is 1.7 km away. There is an elevation difference of 9 m between the two reservoirs. Determine the discharge for the following pipes: (a) commercial steel, (b) cast-iron, (c) aged cast-iron, (d) smooth concrete, and (e) wood-stave pipe in best condition.

3.8.3. It is necessary to deliver 5 l/min of a water–glycerol solution ($d = 1.1$; $v = 1.03 \times 10^{-5}$ m^2/s) under the pressure head of 50 mm Hg. A glass tube is used ($e = 0.003$ mm). Determine the tube diameter if the tube length is 2.5 m.

3.8.4. A lined concrete tunnel of semicircular cross section (radius 1.0 m) has a Hazen–Williams coefficient of 120. What is the head loss if the tunnel carries a discharge of 3 m^3/s over a 1.2-km distance?

3.8.5. Water is taken from a head tank to a cooling pond by a 22-m long, 25-cm diameter steel pipe. If the elevation difference between the tank and the pond is 1.4 m, and if a globe valve is installed in the pipeline, determine the maximum discharge when the valve is fully opened. Assume square-end connections.

3.9   PIPE BRANCHINGS

Water is brought by pipes to a junction where more than two pipes meet. This system must simultaneously satisfy two basic conditions. First, the total amount of water brought by pipes to a junction must always be equal to that carried away from the junction by the other pipes. Second, all pipes that meet at the junction must share the same pressure at the junction.

The hydraulics of pipe branchings at a junction J can be best demonstrated by the classical *three-reservoirs problem*, in which three reservoirs of different elevation are connected to a common junction J as shown in Figure 3.21. Given the lengths, diameters, and materials of all pipes involved, and given the water elevation in each of the three reservoirs, the discharges from or to each reservoir, Q_1, Q_2, and Q_3, can be determined. If an open-ended vertical tube (*piezometer*) is installed at the junction, the water elevation in the tube will rise to the elevation P. The vertical distance between P and J is a direct reading of the pressure head at the junction. The difference in elevations between P and that of reservoir A represents the energy head loss by transporting water from A to J, as indicated by h_{f_1}. The elevation difference between B and P, h_{f_2}, represents the head loss by transporting water from B to J; h_{f_3} represents the head loss by transporting water from J to C.

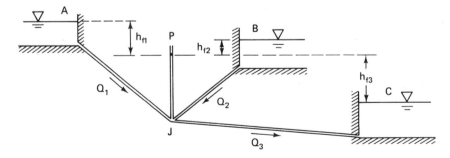

Figure 3.21. Branching pipes connecting three reservoirs.

Since the same quantity of water must be taken away from the junction as is brought to the junction at the same time, we may simply write

$$Q_3 = Q_1 + Q_2 \tag{3.33}$$

or,

$$\Sigma Q = 0, \text{ at the junction.}$$

These types of problems are most conveniently solved by trial. Not knowing the discharge in each pipe, we may first assume a piezometric surface elevation in tube P. This assumed elevation gives the head losses h_{f_1}, h_{f_2}, and h_{f_3} for each of the three pipes. From this set of head losses and the given pipe diameters, lengths, and materials, the trial computation gives a set of values for discharges Q_1, Q_2, and Q_3. If the assumed elevation P is correct, the computed Q's should satisfy the condition stated above, i.e.,

$$\Sigma Q = Q_1 + Q_2 - Q_3 = 0 \tag{3.34}$$

Otherwise, a new elevation P is assumed for the second trial. The computation of another set of Q's is performed until the above condition is satisfied. The correct values of discharge in each pipe are thus obtained.

Note that if the assumed water surface elevation in the piezometer tube is higher than the elevation in reservoir B, Q_1 should be equal to $Q_2 + Q_3$; if it is lower, $Q_1 + Q_2 = Q_3$. The error in the trial Q's indicates the direction in which the assumed piezometric elevation P should be set for the next trial.

It is helpful to plot the computed trial values of P_3 against $(Q_1 + Q_2) - Q_3$. The resulting difference may be either plus or minus for each trial. However, with values obtained from three trials, a curve may be plotted as shown in Figure 3.22. The correct discharge is indicated by the intersection of the curve with the vertical axis. The computation procedure is demonstrated in the following example.

Example 3.10

In Figure 3.22 the three reservoirs A, B, and C are connected by pipes to a common joint J. Pipe AJ is 1000 m long and 30 cm in diameter; pipe BJ is 4000 m long and 50 cm in diameter; and pipe JC is 2000 m long and 40 cm in diameter. The pipes are made of the same material for which $f = 0.03$ may be assumed. Determine the discharge in each pipe. (Neglect minor losses.)

Solution

Let subscripts 1, 2, and 3 represent, respectively, pipes AJ, BJ, and CJ. As a first trial, we assume the pressure head at J, represented by the elevation P, at

Figure 3.22. Trial for three-reservoir problem.

110 m. Q_1, Q_2, and Q_3 may be calculated as follows:

$$h_1 = 120 - 110 = 0.03\left(\frac{L_1}{D_1}\right)\frac{V_1^2}{2g} = 0.03\left(\frac{1000}{0.3}\right)\frac{V_1^2}{2(9.81)}$$

which yields $V_1 = 1.4$ m/s. Hence,

$$Q_1 = V_1 A_1 = (1.4) \cdot \frac{\pi}{4}D_1^2 = \frac{1.4\pi(0.3)^2}{4} = 0.099 \text{ m}^3/\text{s}$$

Similarly,

$$h_2 = 110 - 100 = 0.03\left(\frac{L_2}{D_2}\right)\frac{V_2^2}{2g} = 0.03\left(\frac{4000}{0.5}\right)\frac{V_2^2}{2(9.81)}$$

which yields $V_2 = 0.9$ m/s. Hence,

$$Q_2 = V_2 A_2 = 0.9\left(\frac{\pi}{4}\right)(0.5)^2 = 0.177 \text{ m}^3/\text{s}$$

$$h_3 = 110 - 80 = 0.03\left(\frac{L_3}{D_3}\right)\frac{V_3^2}{2g} = 0.03\left(\frac{2000}{0.4}\right)\frac{V_3^2}{2g}$$

which yields $V_3 = 1.98$ m/s, and

$$Q_3 = V_3 A_3 = 1.98\frac{\pi}{4}(0.4)^2 = 0.249 \text{ m}^3/\text{s}$$

$$Q_1 - (Q_2 + Q_3) = -0.327 \text{ m}^3/\text{s} \neq 0$$

At the second trial, we assume $P = 90$ m. Similar computations are repeated to obtain

$$Q_1 = 0.171 \text{ m}^3/\text{s}$$
$$Q_2 = -0.177 \text{ m}^3/\text{s}$$
$$Q_3 = 0.144 \text{ m}^3/\text{s}$$

Hence,

$$Q_1 - (Q_2 + Q_3) = 0.204 \neq 0.$$

The computations are repeated again by assuming that $P = 85$ m to obtain

$$Q_1 = 0.185 \text{ m}^3/\text{s}$$
$$Q_2 = -0.217 \text{ m}^3/\text{s}$$
$$Q_3 = 0.102 \text{ m}^3/\text{s}$$

Hence,

$$Q_1 - (Q_2 + Q_3) = 0.300 \text{ m}^3/\text{s}.$$

From the above computed values we may make a small plot as shown in Figure 3.22 and find that the elevation $P = 99$ m is where

$$Q_1 - (Q_2 + Q_3) = 0$$

Therefore, $P = 99$ m is taken for the final computation to obtain

$$Q_1 = 0.143 \text{ m}^3/\text{s}$$
$$Q_2 = -0.056 \text{ m}^3/\text{s}$$
$$Q_3 = 0.198 \text{ m}^3/\text{s}.$$

Hence, the condition

$$\Sigma Q = Q_1 - (Q_2 + Q_3) = 0.143 - (-0.056 + 0.198) = 0$$

is met fairly well. The negative value of Q_2 indicates that water flows into reservoir B.

Example 3.11

A horizontal galvanized iron pipe system consists of a 20-cm diameter, 4-m-long main pipe between the two joints 1 and 2. A gate valve is installed at the downstream end immediately before 2. The branch pipe is 12 cm in diameter and 6.4 m long. It consists of two 90° elbows ($R/D = 2.0$) and a globe valve. The system carries a total discharge of 0.26 m³/s water at 10°C. Determine the discharge in each of the pipes when the valves are both fully opened.

Figure 3.23

Solution

The cross-sectional area of pipes a and b are, respectively,

$$A_a = \pi\left(\frac{0.2}{2}\right)^2 = 0.0314 \text{ m}^2; \qquad A_b = \pi\left(\frac{0.12}{2}\right)^2 = 0.0113 \text{ m}^2.$$

Continuity condition requires that

$$0.26 \text{ m}^3/\text{s} = A_a V_a + A_b V_b = 0.0314 V_a + 0.0113 V_b \qquad (3.35)$$

where V_a and V_b are the velocities in pipes a and b, respectively. The head drop between 1 and 2 along the main pipe is h_a:

$$h_a = f_a\left(\frac{L_a}{D_a}\right)\frac{V_a^2}{2g} + 0.15\frac{V_a^2}{2g}$$

The second term is from Table 3.5 for fully opened gate valves. The head drop between 1 and 2 along the branch pipe is h_b:

$$h_b = f_b\left(\frac{L_b}{D_b}\right)\frac{V_b^2}{2g} + 2(0.19)\frac{V_b^2}{2g} + 10\frac{V_b^2}{2g}$$

where the second term is for the elbow losses from Table 3.4 and the third term is for fully opened globe valves from Table 3.5.

Since the pressure head losses through both pipes must be the same, $h_a = h_b$, we have,

$$\left[f_a\left(\frac{4}{0.2}\right) + 0.15\right]\frac{V_a^2}{2g} = \left[f_b\left(\frac{6.4}{0.12}\right) + 0.38 + 10\right]\frac{V_b^2}{2g}$$

or,

$$(20f_a + 0.15)V_a^2 = (53.33f_b + 10.38)V_b^2 \qquad (3.36)$$

Equations (3.35) and (3.36) are solved simultaneously for V_a and V_b. For galvanized iron, Figure 3.7 gives:

$$\left(\frac{e}{D}\right)_a = 0.0006$$

and

$$\left(\frac{e}{D}\right)_b = 0.001$$

Assuming large Reynolds numbers, we may obtain from the Moody diagram the following values:

$$f_a = 0.0175, \text{ and } f_b = 0.0195 \text{ as a first approximation.}$$

Substituting the above values into Eq. (3.36), we have

$$(20 \times 0.0175 + 0.15)V_a^2 = (53.33 \times 0.0195 + 10.38)V_b^2$$
$$0.5V_a^2 = 11.42V_b^2$$
$$V_a = \sqrt{\frac{11.42}{0.5}}\,V_b = 4.78V_b$$

Substituting V_a into Eq. (3.35) we find

$$0.26 = 0.0314\cdot(4.78V_b) + 0.0113V_b = 0.161V_b,$$
$$V_b = \frac{0.26}{0.161} = 1.615 \text{ m/s}$$

Hence, $V_a = 4.78V_b = 4.78 \cdot 1.615 = 7.719$ m/s. Corresponding Reynolds numbers are calculated to check the above friction factors. For pipe a,

$$N_{R_a} = \frac{V_a D_a}{\nu} = \frac{7.719 \cdot 0.2}{1.31 \cdot 10^{-6}} = 1.18 \cdot 10^6$$

The Moody diagram gives $f = 0.0175$ which is correct.
For pipe b,

$$N_{R_b} = \frac{V_b D_b}{\nu} = \frac{1.615 \cdot 0.12}{1.31 \cdot 10^{-6}} = 1.47 \cdot 10^5$$

The Moody diagram gives $f_b = 0.021 \neq 0.0195$.
Equations (3.35) and (3.36) are solved again for the new value of f_b.

$$(20 \cdot 0.0175 + 0.15)V_a^2 = (53.33 \cdot 0.021 + 10.38)V_b^2$$
$$V_a = 4.796 V_b$$

Substituting V_a into Eq. (3.35),

$$0.26 = 0.0314(4.796 V_b) + 0.0113 V_b$$
$$V_b = 1.605 \text{ m/s,} \quad \text{and} \quad V_a = 1.605 \cdot 4.796 = 7.698 \text{ m/s.}$$

Therefore, the discharges are

$$Q_a = A_a V_a = 0.0314 \cdot 7.698 = 0.242 \text{ m}^3/\text{s}$$

and

$$Q_b = A_b V_b = 0.0113 \cdot 1.605 = 0.018 \text{ m}^3/\text{s}$$

Pipe branches connecting more than three reservoirs to a junction are not common in hydraulic engineering. However, problems of multiple (more than three) reservoirs can be solved by following the same principle (see Figure 3.24).

Assume that the water elevation in the piezometer reaches the level P at the junction J. The differences in water surface elevations between reservoirs A and B, A and C, and A and D are, respectively, H_1, H_2, and H_3. The head losses between reservoirs A, B, C, and D and the junction are, respectively, h_{f_1}, h_{f_2}, h_{f_3}, and h_{f_4}, as shown in Figure 3.24. A set of four independent equations may be written in the following general form for the four reservoirs:

$$H_1 = h_{f_1} - h_{f_2} \tag{3.37}$$
$$H_2 = h_{f_1} + h_{f_3} \tag{3.38}$$
$$H_3 = h_{f_1} + h_{f_4} \tag{3.39}$$

In addition, Eq. (3.34) requires that

$$\Sigma Q_j = 0 \tag{3.40}$$

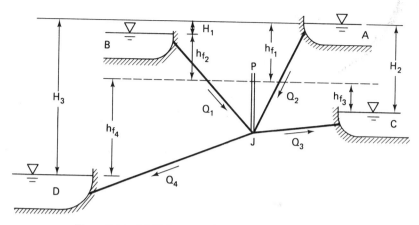

Figure 3.24. Multiple reservoirs connected at a junction.

For each of the pipe branches, the head loss may be expressed in the form of the Darcy–Weisbach equation,* Eq. (3.11), as follows:

$$h_{f_1} = f_1 \left(\frac{L_1}{D_1}\right)\frac{V_1^2}{2g} = f_1 \left(\frac{L_1}{D_1}\right)\frac{Q_1^2}{2gA_1^2}$$

$$h_{f_2} = f_2 \left(\frac{L_2}{D_2}\right)\frac{V_2^2}{2g} = f_2 \left(\frac{L_2}{D_2}\right)\frac{Q_2^2}{2gA_2^2}$$

$$h_{f_3} = f_3 \left(\frac{L_3}{D_3}\right)\frac{V_3^2}{2g} = f_3 \left(\frac{L_3}{D_3}\right)\frac{Q_3^2}{2gA_3^2}$$

$$h_{f_4} = f_4 \left(\frac{L_4}{D_4}\right)\frac{V_4^2}{2g} = f_4 \left(\frac{L_4}{D_4}\right)\frac{Q_4^2}{2gA_4^2}$$

Substituting the above relationship into Eqs. (3.37), (3.38), and (3.39), we have

$$H_1 = \frac{1}{2g} \cdot \left(f_1 \frac{L_1}{D_1}\frac{Q_1^2}{A_1^2} - f_2\frac{L_2}{D_2}\frac{Q_2^2}{A_2^2}\right) \tag{3.41}$$

$$H_2 = \frac{1}{2g} \cdot \left(f_1 \frac{L_1}{D_1}\frac{Q_1^2}{A_1^2} + f_3\frac{L_3}{D_3}\frac{Q_3^2}{A_3^2}\right) \tag{3.42}$$

$$H_3 = \frac{1}{2g} \cdot \left(f_1 \frac{L_1}{D_1}\frac{Q_1^2}{A_1^2} + f_4\frac{L_4}{D_4}\frac{Q_4^2}{A_4^2}\right) \tag{3.43}$$

and $$\Sigma Q_j = 0. \tag{3.40}$$

---

*Other empirical formulas, such as the Manning formula and the Hazen–Williams formula, may be applied using corresponding expressions.

Equations (3.41), (3.42), (3.43), and (3.40) can then be solved simultaneously for the four unknowns Q_1, Q_2, Q_3, and Q_4. These values are the discharge for each of the pipe branches shown. This procedure could be applied to any number of reservoirs connected to a common junction.

PROBLEMS

3.9.1. Determine the flow into or out of each reservoir in Figure 3.25 if the connecting pipes all have a Hazen–Williams coefficient of 120.

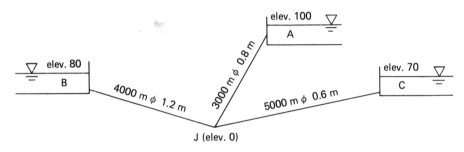

Figure 3.25

3.9.2. Determine the flow into or out of each reservoir in Figure 3.25 if the connecting pipes are made of the same material with $e = 0.05$ mm and water temperature at 20°C.

3.9.3. A 1.3-m concrete pipe 2000 m long carries water at 5 m³/s from reservoir A to a junction and discharges the water into reservoir B and reservoir C. All connecting pipes are 1 m in diameter. The distances from the junction to reservoirs B and C are each 1500 m, and the elevation difference between A and B is 8 m. Determine the surface elevation at C.

3.10 PIPE NETWORKS

The water supply distribution system in a municipal district is usually constructed of a large number of pipes connected together to form loops and branches. Although the calculations of water flow in a network involve a large number of pipes and may become tedious, the solution to the problem is based on the same principles that govern the pipelines and pipe branchings previously discussed. In general, a series of simultaneous equations can be written for the network. These equations are written to satisfy the following conditions:

1. At any junction, $\Sigma Q = 0$ (junction equation).
2. Between any two junctions the total head loss is independent of the path taken (loop equation).

Depending on the number of unknowns, it is usually possible to set up a sufficient number of independent equations to solve the problem. A typical problem would, for example, be to determine the flow distribution in each pipe of a network when the total inflows (say, Q_1 and Q_2) and outflows (say, Q_3 and Q_4) are given. These equations may then be solved simultaneously.

For the simple network shown in Figure 3.26, a set of 12 independent equations (eight junction equations and four loop equations) is needed to solve for the flow distribution in the 12 pipes. As a general rule, a network with m loops and n junctions provides a total of $m + (n - 1)$ independent equations. For more complex networks, the number of equations increases proportionally. At a certain point it will be obvious that algebraic solution of the network equations becomes impractical. In such a case, the solutions may feasibly be obtained by means of programming the equations for a digital computer. A commonly used computer program for a pipe network is developed from the *Hardy–Cross method*, which is outlined below.

The Hardy–Cross method is a method of successive approximations based on the two conditions stated above for each junction and loop in the network system. In the elemental loop A shown in Figure 3.26, for example, the arrowheads indicate the estimated direction of flow. This loop must satisfy the following conditions: (1) At each junction b, c, d, or e the total inflow must equal the total outflow. (2) The loss of head due to flow in the counterclockwise direction along pipes bc and cd must be equal to that of flow in the clockwise direction along pipes be and ed.

As a first approach, a flow distribution that meets condition (1) is assumed. The chance for such an assumed flow distribution to meet condition (2) is

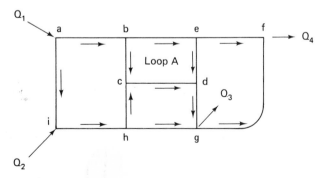

Figure 3.26

small because the computed head loss in pipes along the clockwise direction, based on the assumed flow rates, may not ordinarily be equal to that along the counterclockwise direction. The successive computational procedure corrects the assumed flows and tends to equalize the head loss.

Knowing the diameter, length, and roughness of a pipe, we see that the head loss in the pipe is a function of the flow rate Q. Applying the Darcy–Weisbach formula [Eq. (3.4)], we may write

$$h_f = f\left(\frac{L}{D}\right)\frac{V^2}{2g} = \left[f\left(\frac{L}{D}\right) \cdot \frac{1}{2gA^2}\right] \cdot Q^2 = KQ^2 \qquad (3.44)$$

In any elementary loop, such as loop A, the total head loss in the clockwise direction (subscript c) is the sum of head losses in all pipes that carry flow in the clockwise direction around the loop.

$$\Sigma h_{f_c} = \Sigma K_c Q_c^2 \qquad (3.45)$$

Similarly, the loss of head in the counterclockwise direction (subscript cc) is

$$\Sigma h_{f_{cc}} = \Sigma K_{cc} Q_{cc}^2 \qquad (3.46)$$

With the assumed flow rates, Q's, it is not expected that these two values will be equal during the first trial, as mentioned previously. The difference,

$$\Sigma K_c Q_c^2 - K_{cc} Q_{cc}^2$$

is the *closure error* of the first trial.

It is desired to determine a flow correction ΔQ which, when subtracted from Q_c and added to Q_{cc}, will equalize the two head losses. Thus, the correction ΔQ must satisfy the following equation:

$$\Sigma K_c (Q_c - \Delta Q)^2 = \Sigma K_{cc}(Q_{cc} + \Delta Q)^2$$

Expanding the terms in the parentheses on both sides of the equation, we have

$$\Sigma K_c (Q_c^2 - 2Q_c\Delta Q + \Delta Q^2) = \Sigma K_{cc}(Q_{cc}^2 + 2Q_{cc}\Delta Q + \Delta Q^2)$$

Assuming that the correction term is small compared to both Q_c and Q_{cc}, we may simplify the above expression by dropping the last term on each side of the equation and write

$$\Sigma K_c (Q_c^2 - 2Q_c\Delta Q) = K_{cc}(Q_{cc}^2 - 2Q_c\Delta Q)$$

From this relationship we may solve for ΔQ:

$$\Delta Q = \frac{\Sigma K_c Q_c^2 - \Sigma K_{cc} Q_{cc}^2}{2(\Sigma K_c Q_c + \Sigma K_{cc} Q_{cc})} \qquad (3.47)$$

If we take Eq. (3.44) and divide it by Q on both sides, we have

$$KQ = \frac{h_f}{Q} \qquad (3.48)$$

Equations (3.45), (3.46), and (3.48) can be substituted into Eq. (3.47) to obtain

$$\Delta Q = \frac{\Sigma h_{f_c} - \Sigma h_{f_{cc}}}{2\left(\Sigma \dfrac{h_{f_c}}{Q_c} + \Sigma \dfrac{h_{f_{cc}}}{Q_{cc}}\right)}{}^{*} \qquad (3.49)$$

The second approximation involves this correction for a new estimated flow rate distribution. The computation results from the second approximation are expected to give a closer match of the two head losses around the c and cc directions in loop A. Note that pipes bc, cd, and ed in loop A are each common to two loops and, therefore, are subjected to double corrections from each loop. The successive computation procedure is repeated until each loop in the entire network is relaxed and the corrections become negligibly small.

The Hardy–Cross method can be conveniently handled by the digital computer. A typical program and an example problem of application are provided in Figure 3.27.

Example 3.12

Determine the discharge in each pipe in the pipe network shown in Figure 3.27. If the pressure head measured at junction 1 is 100 m, determine the pressure head at each of the junctions in the network. All pipes are new cast-iron and are in good condition.

Solution

The Hardy–Cross method is used to solve this problem by using the computer program.

The computer program, as shown in Figure 3.28(a), uses the Hazen–Williams formula [Eq. (3.22)] to compute the head loss in each pipe based

*When the Hazen–Williams formula (Eq. 3.22) is used, the equation should be,

$$\Delta Q = \frac{\Sigma h_{f_c} - \Sigma h_{f_{cc}}}{1.85\left(\Sigma \dfrac{h_{f_c}}{Q_c} + \Sigma \dfrac{h_{f_{cc}}}{Q_{cc}}\right)} \qquad (3.49a)$$

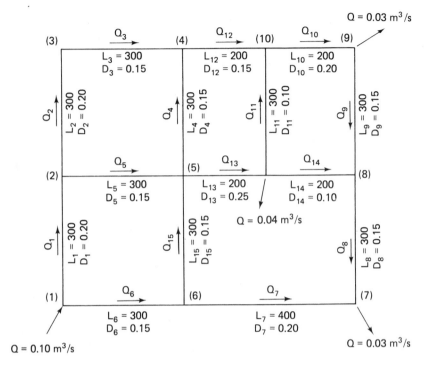

Figure 3.27

on the assumed flow rate. The Hardy–Cross method is used to balance the flow rate-pressure relationship in each loop. The program is provided for general application of any numbers and combinations of loops and junctions in a network of given pipe material (Hazen–Williams coefficients).

For the first approximation, a set of discharges is arbitrarily assumed to satisfy the continuity condition at each joint, i.e., $\Sigma Q = 0$.

$$
\begin{array}{llll}
Q_1 = 0.05 & Q_2 = 0.03 & Q_3 = 0.03 & Q_4 = 0.00 \\
Q_5 = 0.02 & Q_6 = 0.05 & Q_7 = 0.02 & Q_8 = 0.01 \\
Q_9 = 0.01 & Q_{10} = 0.02 & Q_{11} = 0.01 & Q_{12} = 0.03 \\
Q_{13} = 0.05 & Q_{14} = 0.02 & Q_{15} = 0.03 &
\end{array}
$$

The program shows that five consecutive iterations were made in the computation. The results are shown in Figure 3.28(b).

```
C     THIS PROGRAM READS AN ARBITRARY PIPE NETWORK WITH ESTIMATED FLOWS.
C     THE HAZEN WILLIAMS FORMULA IS USED TO COMPUTE THE HEAD LOSSES FOR
C     EACH PIPE. EACH LOOP IS BALANCED BY THE HARDY CROSS METHOD.
C     PRESSURES AT EACH JUNCTION ARE CALCULATED BASED ON THE PRESSURE AT
C     A GIVEN JUNCTION AND THE ELEVATION OF EACH JUNCTION.
C
      DIMENSION LOOP(15,10),NR(15),P(100),Q(100),JI(100),JF(100),IR(100)
     1,FRIC(100),TE(100),HLOSS(100)
      REAL LGTH
      DO 10 I=1,100
   10 P(I)=999.
C
C     **READ NETWORK DATA. NUMBER OF REACHES AND JUNCTIONS,TOLERANCES ETC.
C
      READ(5,12)NRS,NODES,LOOPS,ITER,TOLQ,TOLH,PKNOW,NODE
   12 FORMAT(4I5,3F10.0,I5)
      WRITE(6,13)NRS,NODES,LOOPS,ITER,TOLQ,TOLH,NODE,PKNOW
   13 FORMAT(2X,'SOLUTION OF A PIPE NETWORK BY THE HARDY CROSS METHOD'//
     1/2X,'DATA INPUT'//5X,'NUMBER OF REACHES =',I3/5X,'NUMBER OF NODES
     2    =',I3/5X,'NUMBER OF LOOPS    =',I3/5X,'ITERATION LIMIT     =',I
     33/5X,'DISCHARGE TOLERANCE=',F7.3/5X,'HEAD LOSS TOLERANCE=',F7.3/5X
     4,'PRESSURE AT JUNC.',I2,'=',F6.2/)
C
C     **READ LOOPS AND CORRESPONDENT REACHES
C
      DO 16 NL=1,LOOPS
      READ(5,14)JZ,(LOOP(NL,I),I=1,JZ)
   14 FORMAT(11I5)
      WRITE(6,15)JZ,NL,(LOOP(NL,I),I=1,JZ)
   15 FORMAT(5X,I2,' REACHES IN LOOP',I3,' ———',10I5)
   16 NR(NL)=JZ
```

Figure 3.28(a). A typical program for pipe network computation.

```fortran
32          P(NODE)=PKNOW
33          DO 17 IP=1,NRS
34   17     Q(IP)=0.
35   C
36   C      **READ THE DESCRIPTION OF THE REACHES,NUMBER,INITIAL AND FINAL JUNCTION...
37   C
38          WRITE(6,18)
39   18     FORMAT(/5X,'REACHES INFORMATION'/7X,'REACH      IJUNC     FJUNC      L(M)
40   1          D(M)      Q(CMS)          C')
41          DO 20 I=1,NRS
42          READ(5,58)L,JI(L),JF(L),C,LGTH,D,Q(L)
43   58     FORMAT(3I5,F5.0,3F10.0,I5)
44          WRITE(6,19)L,JI(L),JF(L),LGTH,D,Q(L),C
45   19     FORMAT(3X,3I8,F10.2,F9.2,F11.3,F8.0)
46          IR(I)=L
47   20     FRIC(L)=10.63*LGTH/((C**1.85)*(D**4.87))
48   C
49   C      **BEGIN BALANCING PROCEDURE
50   C
51          DO 32 IT=1,ITER
52          LOK=0
53          DO 31 K=1,LOOPS
54          SUMH=0.
55          SUMZ=0.
56          IWX=NR(K)
57   C
58   C      **CALCULATE THE CORRECTION FACTOR FOR EACH LOOP
59   C
60          DO 25 IRL=1,IWX
61          S1=1
62          S2=1
```

Figure 3.28(a). Cont.

```
63      L=LOOP(K,IRL)
64      IF(L)21,25,22
65  21  S1=-1
66      L=-L
67  22  FLOWR=Q(L)
68      IF(FLOWR)23,24,24
69  23  S2=-1
70      FLOWR=-FLOWR
71  24  H=S1*S2*FRIC(L)*FLOWR**1.85
72      Z=1.85*FRIC(L)*FLOWR**0.85
73      SUMH=SUMH+H
74      SUMZ=SUMZ+Z
75  25  CONTINUE
76      FCORR=SUMH/SUMZ
77  C
78  C   **CORRECT DISCHARGES FOR THIS LOOP
79  C
80      DO 28 TRL=1,TWX
81      L=LOOP(K,IRL)
82      IF(L)26,28,27
83  26  L=-L
84      Q(L)=Q(L)+FCORR
85      GO TO 28
86  27  Q(L)=Q(L)-FCORR
87  28  CONTINUE
88  C
89  C   **TEST CORRECTION FACTORS AGAINST TOLERANCES
90  C
91      TEST=ABS(FCORR)-TOLQ
92      IF(TEST)30,30,29
93  29  TEST=ABS(SUMH)-TOLH
```

Figure 3.28(a). Cont.

```
94          IF(TEST)30,30,31
95     30   LOK=LOK+1
96     31   CONTINUE
97          IF(LOOPS .LE. LOK)GO TO 33
98     32   CONTINUE
99    C
100   C      **READ THE TOPOGRAPHY AND CALCULATE THE PRESSURES
101    33   WRITE(6,34)
102    34   FORMAT(/5X,'JUNCTIONS INFORMATION'/7X,'JUNC. ELEVATION')
103          DO 36 I=1,NODES
104          READ(5,35)N,TE(N)
105    35   FORMAT(I5,F10.0)
106    36   WRITE(6,37)N,TE(N)
107    37   FORMAT(7X,I3,4X,F8.4)
108          DO 38 I=1,NRS
109          L=IR(I)
110          S=1
111          IF(Q(L) .LT. 0.0)S=-1
112    38   HLOSS(L)=S*FRIC(L)*(ABS(Q(L)))**1.85
113    39   IFLAG=0
114          DO 50 K=1,NRS
115          I=IR(K)
116          JUPS=JI(I)
117          JDOWN=JF(I)
118          IF(P(JUPS)-999.)40,42,42
119    40   IF(P(JDOWN)-999.)50,41,41
120    41   P(JDOWN)=P(JUPS)-HLOSS(I)+TE(JUPS)-TE(JDOWN)
121          GO TO 50
122    42   IF(P(JDOWN)-999.)44,43,43
123    43   IFLAG=1
124          GO TO 50
```

Figure 3.28(a). Cont.

```
125   44   P(JUPS)=P(JDOWN)+HLOSS(I)-TE(JUPS)+TE(JDOWN)
126   50   CONTINUE
127        IF(IFLAG .EQ. 1)GO TO 39
128        WRITE(6,51)IT
129   51   FORMAT(//5X,'FLOWS AFTER',I3,' ITERATIONS ARE'/7X,'REACH        Q(CMS)'
130        1)
131        DO 52 I=1,NRS
132   52   WRITE(6,53)I,Q(I)
133   53   FORMAT(8X,I3,4X,F6.3)
134        WRITE(6,54)
135   54   FORMAT(/5X,'PRESSURE AT EACH JUNCTION IN COLUMN OF WATER'/8X,'JUNC
136        1.   P(M)')
137        DO 55 I=1,NODES
138   55   WRITE(6,56)I,P(I)
139   56   FORMAT(8X,I3,3X,F7.3)
140   99   STOP
141        END
```

Figure 3.28(a). Cont.

95

```
DATA INPUT
   NUMBER OF REACHES  = 15
   NUMBER OF NODES    = 11
   NUMBER OF LOOPS    =  5
   ITERATION LIMIT       = 50
   DISCHARGE TOLERANCE=  0.001
   HEAD LOSS TOLERANCE=  0.005
   PRESSURE AT JUNC. 1 =100.00
   4 REACHES IN LOOP  1 ──────   1    5  -15   -6
   4 REACHES IN LOOP  2 ──────   2    3   -4   -5
   4 REACHES IN LOOP  3 ──────   4   12   11  -13
   4 REACHES IN LOOP  4 ────── -11   10   -9  -14
   5 REACHES IN LOOP  5 ──────  15   13   14    8   -7
   REACHES INFORMATION
```

REACH	IJUNC	FJUNC	L(M)	D(M)	Q(CMS)	C
1	1	2	300.00	0.20	0.050	130.
2	2	3	300.00	0.20	0.030	130.
3	3	4	300.00	0.15	0.030	130.
4	5	4	300.00	0.15	0.	130.
5	2	5	300.00	0.15	0.020	130.
6	1	6	300.00	0.15	0.050	130.
7	6	7	400.00	0.20	0.020	130.
8	8	7	300.00	0.15	0.010	130.
9	8	9	300.00	0.15	0.010	130.
10	10	9	200.00	0.20	0.020	130.
11	10	11	300.00	0.10	0.010	130.
12	4	10	200.00	0.15	0.030	130.
13	5	11	200.00	0.25	0.050	130.
14	11	8	200.00	0.10	0.020	130.
15	6	5	300.00	0.15	0.030	130.

```
JUNCTIONS INFORMATION
   JUNC. ELEVATION
```

JUNC.	ELEVATION
1	0.
2	0.
3	0.
4	0.
5	0.
6	0.
7	0.
8	0.
9	0.
10	0.
11	0.

```
FLOWS AFTER  5 ITERATIONS ARE
   REACH  Q(CMS)
```

REACH	Q(CMS)
1	0.059
2	0.027
3	0.027
4	-0.007
5	0.031

Figure 3.28(b). Solution of a pipe network by the Hardy Cross method

6	0.041
7	0.032
8	−0.002
9	0.006
10	0.024
11	−0.003
12	0.021
13	0.047
14	0.004
15	0.009

PRESSURE AT EACH JUNCTION IN COLUMN OF WATER

JUNC.	P(M)
1	100.000
2	94.770
3	93.506
4	88.378
5	88.007
6	88.902
7	86.632
8	86.592
9	86.250
10	86.902
11	87.549

Figure 3.28(b). Cont.

PROBLEMS

3.10.1. The total discharge from A to B in Figure 3.29 is 12 l/s. Determine the head loss between A and B and the flow rate in each pipe if cast-iron pipes are used.

Figure 3.29

3.10.2. The system in Figure 3.30 carries a total discharge of 1.3 m³/s. All pipes are commercial steel. Determine the head loss between A and D.

Figure 3.30

3.10.3. Parallel cast-iron pipes 1, 2, and 3 in Figure 3.31 carry a total discharge of 0.8 m³/s. Determine the flow rate in each pipe. Neglect minor losses.

Figure 3.31

3.10.4. Determine the flow in each pipe in the network shown in Figure 3.32.

$Q_1 = 1.0$ m³/sec.
$Q_2 = 3.0$ m³/sec.
$Q_3 = 2.5$ m³/sec.
$Q_4 = 0.25$ m³/sec
$Q_5 = 1.25$ m³/sec.

Figure 3.32

3.11   PIPELINE WITH NEGATIVE PRESSURE OR PUMPS

Long pipelines laid to transport water from one reservoir to another over a large distance usually follow the natural configuration of the land. Occasionally, a section of the piepline may be raised to an elevation that is above the local hydraulic gradient line, as shown in Figure 3.33. As discussed earlier in this chapter, the vertical distance measured between the energy gradient line and the hydraulic gradient line at any location along the pipeline is the velocity head, $V^2/2g$, at that location; and the vertical distance measured between the hydraulic gradient line and the pipeline is the local pressure head, P/γ. In the vicinity of the pipeline summit, S, the pressure head may take on negative values. This is because the total energy head

$$H = \frac{V^2}{2g} + \frac{P}{\gamma} + h$$

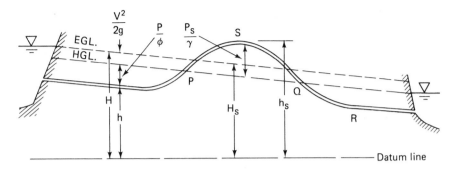

Figure 3.33. Elevated section in a pipeline.

must equal the vertical distance between the datum and the energy gradient line at any location in the pipeline. At the summit, for example, the elevation h_s is given and the velocity head $V^2/2g$ is also a fixed positive value. The sum $V^2/2g + h_s$ may become larger than H_s. Hence, the pressure head, P_s/γ_1, can only be negative. Negative pressure (reference to atmospheric pressure as zero, $P_{atm} = 0$) exists in the pipeline wherever the pipeline is raised above the hydraulic gradient line (between P and Q). This negative pressure reaches a maximum value at the summit, $-(P_s/\gamma)$. Water flow from S to R must flow against the pressure gradient. In other words, it flows from a point of lower pressure toward a point of higher pressure. This is possible because water always flows from high-energy toward lower-energy locations, and in the closed conduit the elevation head converts to compensate the pressure head along the gradient line. For example, if a unit weight of water flowing from S to R experiences a pressure increase equal to 3 m of water column, the elevation of S must be at least 3 m higher than the elevation of R. The difference in elevation between S and R must equal exactly 3 m plus the loss of head between S and R. Or, more generally,

$$h_{12} = \left(\frac{P_2}{\gamma} - \frac{P_1}{\gamma}\right) + h_{f_{12}} \qquad (3.50)$$

between any two points 1 and 2 in a pipeline.

It is, however, important to maintain pressure in a pipeline everywhere above the vapor pressure of water. As we have discussed in Chapter 1, the vapor pressure of water is approximately equal to a negative water column of 10 m at 20°C. When the pressure in a pipe drops below this value, water will be vaporized locally to form vapor pockets that separate the water in the pipe. These vapor pockets collapse in regions of higher pressure downstream. The action of vapor collapse is most violent, causing vibrations and sound that can do great damage to the pipeline.

Theoretically, a pipeline may be designed to allow pressure to fall to the level of vapor pressure at certain sections in the pipeline. In practice, however, most of the time water contains dissolved gases that will come out from the water solution well before the vapor pressure point is reached. Such gases return into the solution very slowly. They usually move with the water in the form of large bubbles that reduce the pipe flow cross section and tend to disrupt the flow. For this reason, negative pressure should be kept from exceeding approximately two-thirds of barometric head in any section of the pipeline.

Example 3.13

A uniform steel pipeline 40 cm in diameter and 2000 m long carries water at 10°C between two reservoirs as shown in Figure 3.33. The two reservoirs have a water surface elevation difference of 30 m. At midlength the pipeline must be raised to carry the water over a small hill. Determine the maximum elevation to which the summit S may be raised above the lower reservoir water.

Solution

The total head loss between the two reservoirs is 30 m, which includes the entrance loss, h_e, the friction loss in the pipeline, h_f, and the discharge loss, h_d. We may write

$$30 = \left(K_e + f\frac{L}{D} + K_d\right)\frac{V^2}{2g}$$

From Eqs. (3.27), (3.11), (3.30) assuming square entrance and that $f = 0.025$, we have

$$30 = \left(0.5 + 0.025 \cdot \frac{2000}{0.4} + 1\right)\frac{V^2}{2(9.81)}$$

Hence, $V = 2.157$ m/s. We must check the assumed friction factor f against this velocity. For commercial steel, $e/D = 0.00011$, and the Reynolds number is

$$N_R = \frac{VD}{\nu} = \frac{2.157 \cdot 0.4}{1.31 \cdot 10^{-6}} = 6.6 \cdot 10^5$$

The Moody friction chart (Figure 3.8) gives $f = 0.014$, which is significantly different from the assumed value. The energy relationship must be rewritten as follows:

$$30 = \left(0.5 + 0.014 \cdot \frac{2000}{0.4} + 1\right)\frac{V^2}{2(9.81)}$$

From which $V = 2.87$ m/s. Here, $N_R = 8.76 \cdot 10^5$, and the Moody friction chart gives $f = 0.0138$, which is close enough to the second approximation, and the velocity of 2.87 m/s is accepted for the following computations. At 10°C the water vapor pressure head is 10.2 m. The maximum elevation that the pipeline summit, S, may be raised, h_s, is therefore

$$h_s = (0 + 10.2) + \left(1 + f\frac{L}{D}\right)\frac{V^2}{2g}$$

$$= 10.2 + \left[1 + (0.014)\frac{1000}{0.4}\right]\frac{2.87^2}{2(9.81)} = 25.3 \text{ m}$$

Pumps may be needed in a pipeline to lift water from a lower elevation or simply to boost the rate of flow. Pump operation adds energy to water in the pipeline by boosting the pressure head. The details of pump design and selection will be discussed in Chapter 4, but the analysis of pressure head and energy head provided by a pump to a pipeline system are demonstrated in the following example.

Example 3.14

An emergency pump is installed in the pipeline system of Example 3.13 at a distance of 500 m from reservoir A. The pump is used to boost the flow rate when needed. Determine the pressure head the pump must develop in order to double the flow rate.

Solution

The head loss between reservoir A and the pump is

$$h_{f_{AP}} = \left(0.5 + 0.014 \cdot \frac{500}{0.4}\right)\frac{V^2}{2g} \tag{1}$$

The head loss between the pump and reservoir B is

$$h_{f_{PB}} = \left(0.014 \cdot \frac{1500}{0.4} + 1\right)\frac{V^2}{2g} \tag{2}$$

The rate of flow in the pipeline without the pump, from Example 3.8, is

$$Q_0 = AV = \frac{\pi}{4}(0.4)^2(2.87) = 0.36 \text{ m}^3/\text{s}$$

The pump operation will double the flow rate:

$$Q = 2Q_0 = 2 \cdot 0.36 = 0.72 \text{ m}^3/\text{s}$$

and the corresponding velocity is

$$V = \frac{Q}{A} = 5.73 \text{ m/s}$$

Substituting this value into Eqs. (1) and (2), we have

$$h_{f_{AP}} = \left(0.5 + 0.014 \cdot \frac{500}{0.4}\right)\frac{5.73^2}{2(9.81)} = 30.12 \text{ m}$$

$$h_{f_{PB}} = \left(0.014 \cdot \frac{1500}{0.4} + 1\right)\frac{5.73^2}{2(9.81)} = 89.53 \text{ m}$$

The minimum pressure head the pump must provide to double the flow rate is

$$H_P = 30.12 + 89.53 - 30 = 89.65 \text{ m}$$

In addition to this value, a certain amount of pressure head must also be provided to compensate the energy loss that occurs at the pump when it is in operation.

PROBLEMS

3.11.1. A 10-cm tube is 12 m long and is used to siphon water from the reservoir and discharge it into the air, as shown in Figure 3.34. If the total head loss between the intake end of the tube and the summit, S, is 1.2 m and that between S and the discharge end is 2.0 m, determine the discharge of the tube and the pressure at S.

Figure 3.34

3.11.2. A siphon spillway with a square cross section of 1 m² is used to discharge water to a downstream reservoir 20 m below the crest of the spillway (see Figure 3.35). If the frictional head loss is equal to twice the velocity head and is evenly distributed throughout the spillway length, determine the range of water surface level within which the spillway may operate. The crest of the spillway is located at one-third length from the entrance.

Figure 3.35

3.11.3. A pump draws water from reservoir *A* and lifts it to a higher reservoir *B*, as shown in Figure 3.36. The head loss from *A* to the pump is four times the velocity head in the 10-cm pipe, and the head loss from the pump to *B* is seven times the velocity head. Compute the pressure head the pump must deliver. The pressure head at the inlet of the pump is -6 m.

Figure 3.36

3.11.4. Construct the energy gradient line and the hydraulic gradient line for the pipeline system in Prob. 3.11.3.

3.12   WATER HAMMER PHENOMENON IN PIPELINES

A sudden change of flow rate in a large pipeline (due to valve closure, pump turnoff, etc.) may involve a great mass of water moving within the pipe walls. The force resulting from changing the speed of the water mass may cause a pressure rise in the pipe with a magnitude several times greater than the normal static pressure in the pipe. This phenomenon is commonly known as the *water hammer phenomenon*. The excessive pressure may fracture the pipe walls or cause other damage to the pipeline system. The possible occurrence of such pressure, its magnitude, and the propagation of the pressure wave form must be carefully investigated in connection with the pipeline design.

The sudden change of pressure due to a valve closure may be viewed as the result of the force developed in the pipe necessary to stop the flowing water column. The column has a total mass M and is changing its velocity at the rate of dV/dt. According to Newton's second law of motion,

$$F = m\frac{dV}{dt} \tag{3.51}$$

If the velocity of the entire water column could be reduced to zero instantly, the above equation would become

$$F = \frac{m(V_0 - 0)}{0} = \frac{mV_0}{0} = \infty$$

The resulting force (hence, pressure) would be infinite. Fortunately, such an instantaneous change is almost impossible because a mechanical valve requires a certain amount of time to complete a closure operation. In addition, neither the pipe walls nor the water column involved are perfectly rigid under large pressure. The elasticities of both the pipe walls and the water column play very important roles in the water hammer phenomenon.

Consider a pipe of length L with inside diameter D, wall thickness t, and the modulus of elasticity E_p. Immediately following the valve closure, the water in close proximity to the valve is brought to rest, and its kinetic energy head, $V^2/2g$, is transformed to pressure energy head, P/γ, and causes a local pressure increase. As a result of this pressure increase, the water column in this section is somewhat compressed, and the pipe walls also expand slightly due to the corresponding increase of stress in the walls. Both of these phenomena help provide a little extra volume, allowing water to enter the section continuously until it comes to a complete stop.

The next section immediately upstream is involved in the same procedure an instant later. In this manner, a wave of increased pressure propagates up the pipe toward the reservoir, as shown in Figure 3.37(b). When this pressure wave reaches the upstream reservoir, the entire pipe is expanded and the water column within is compressed by the increased pressure. At this very instant the entire water column within the pipe comes to a complete halt.

Obviously, this transient state cannot be maintained because the pressure in the pipe is much higher than the pressure in the open reservoir. The halted water in the pipe begins to flow back into the reservoir as soon as the pressure

Figure 3.37. Propagation of water hammer pressure waves (friction in pipe neglected). (a) Steady-state condition previous to valve movement. (b) Transient conditions at $t < L/C$. (c) Transient conditions at $t = L/C$. (d) Transient conditions at $L/C < t < 2L/C$. (e) Transient conditions at $t = 2L/C$. (f) Transient conditions at $2L/C < t < 3L/C$. (g) Transient conditions at $t = 3L/C$. (h) Transient conditions at $4L/C > t > 3L/C$. (i) Transient conditions at $t = 4L/C$. Notes: After $t = 4L/C$ the cycle repeats and continues indefinitely if the friction in the pipe is zero. The symbol ↄ or ↄ is used to denote the reflection of the wave front.

wave reaches the reservoir. This process starts at the reservoir end of the pipe and a decreased pressure wave travels downstream toward the valve, as shown in Figure 3.37(d). During this period the water behind the wave front moves in the upstream direction as the pipe continuously contracts and the column decompresses. The time required for the pressure wave to return to the valve is $2L/C$, where C is the speed of the wave travel along the pipe. It is also known as *celerity*.

The speed of pressure wave travel in a pipe depends on the modulus of elasticity of water E_b, and the modulus of elasticity of the pipe wall material E_p. The relationship may be expressed as

$$C = \sqrt{\frac{E_c}{\rho}} \qquad (3.52)$$

where E_c is the composite modulus of elasticity of the water pipe system and ρ is the density of water. E_c is composed of the elasticity of the pipe walls and the elasticity of the fluid within. It may be calculated by the following relationship:

$$\frac{1}{E_c} = \frac{1}{E_b} + \frac{Dk}{E_p t} \qquad (3.53)$$

The modulus of elasticity of water, E_b, and the density of water are given in Chapter 1. The modulus of elasticity of various common pipe materials is listed in Table 3.6, k is a constant depending upon the method of pipeline

TABLE 3.6 Modulus of Elasticity, E_p, of Common Pipe Materials

Pipe Material	E_p (dyn/cm²)	(psi)
Aluminum	$7 \cdot 10^{11}$	$10 \cdot 10^6$
Brass, Bronze	$9 \cdot 10^{11}$	$13 \cdot 10^6$
Cast-iron, gray	$11 \cdot 10^{11}$	$16 \cdot 10^6$
Cast-iron, malleable	$16 \cdot 10^{11}$	$23 \cdot 10^6$
Concrete, reinforced	$16 \cdot 10^{11}$	$25 \cdot 10^6$
Glass	$7 \cdot 10^{11}$	$10 \cdot 10^6$
Lead	$31 \cdot 10^8$	$4.5 \cdot 10^4$
Lucite	$28 \cdot 10^8$	$4 \cdot 10^4$
Copper	$97 \cdot 10^{10}$	$14 \cdot 10^6$
Rubber, vulcanized	$14 \cdot 10^{10}$	$2 \cdot 10^6$
Steel	$19 \cdot 10^{11}$	$28 \cdot 10^6$

anchoring, and t is the thickness of the pipe walls.

$k = \left(\dfrac{5}{4} - \epsilon\right)$, for pipes free to move longitudinally;

$k = (1 - \epsilon^2)$, for pipes anchored at both ends against longitudinal movement;

$k = (1 - 0.5\epsilon)$, for pipes with expansion joints,

where ϵ is the Poisson's ratio of the pipe wall material. It may take on the value $\epsilon = 0.25$ for common pipe materials.

If the longitudinal stress in a pipe can be neglected, $k = 1.0$, and Eq. (3.53) can be simplified:

$$\frac{1}{E_c} = \frac{1}{E_b} + \frac{D}{E_p t} \tag{3.53a}$$

Figure 3.37(e) shows that by the time the decreased pressure wave arrives at the valve, the entire column of water within the pipe is in motion in the upstream direction. This motion cannot pass the already closed valve and must be stopped when the wave arrives at the valve. The inertia of this moving water mass causes the pressure at the valve to drop below the normal static pressure. A third oscillation period begins as the wave of negative pressure propagates up the pipe toward the reservoir, as shown in Figure 3.37(f). At the instant the negative pressure reaches the reservoir the water column within the pipe again comes to a complete standstill and the pressure in the pipe is less than the static pressure. Since the pressure at the entrance to the pipe is less than that in the reservoir, water flows again into the pipe, starting a fourth period of oscillation.

The fourth period is marked by a wave of normal static pressure moving downstream toward the valve, as in Figure 3.37(h). The water mass behind the wave front also moves in the downstream direction. This fourth-period wave arrives at the valve at time $4L/C$, with the entire pipe returning to normal static pressure and water moving in the downstream direction. For an instant the conditions throughout the pipe are somewhat similar to that at the time of the valve closure (the beginning of the first-period wave), except that the water velocity in the pipe is less than it was at the time of valve closure. This is a result of energy losses to heat due to friction and the viscoelastic behavior of the pipe walls and the water column.

Another cycle begins instantly. The four sequential waves travel up and down the pipe in exactly the same manner as the first cycle described above, except that the corresponding pressure waves are much smaller in magnitude. The pressure wave oscillation continues with each set of waves successively diminishing, until finally the waves die out.

As mentioned previously, the closure of a valve usually requires a certain period of time t to complete. If t is less than $2L/C$ or if the valve closure is completed before the first pressure wave could return to the valve, the resulting rise in pressure should be the same as if the valve had an instantaneous closure. However, if t is greater than $2L/C$, then the first pressure wave returns to the valve before the valve is completely closed. The returned negative pressure wave compensates the pressure rise resulting from the final closure of the valve.

The maximum pressure created by the water hammer phenomenon may be calculated. The formulas for water hammer pressure are derived as follows.

Consider a pipe with a rapidly closing valve ($t \leq 2L/C$). The extra volume of water ΔVol that enters the pipe during the first period ($t = -L/C$) [Figure 3.37(c)] is

$$\Delta\text{Vol} = -V_0 A\left(\frac{L}{C}\right) \tag{3.54}$$

where V_0 is the initial velocity of water flowing in the pipe and A is the pipe cross-sectional area. The increased pressure ΔP is related to this extra volume by

$$\Delta P = -E_c \cdot \frac{\Delta\text{Vol}}{\text{Vol}} = -\frac{E_c \cdot \Delta\text{Vol}}{AL} \tag{3.55}$$

where Vol is the original volume of the water column in the pipe and E_c is the combined modulus of elasticity, as defined by Eq. (3.53). Substituting Eq. (3.54) into Eq. (3.55), we may write

$$\Delta P = \frac{E_c}{AL} \cdot V_0 \cdot A \cdot \left(\frac{L}{C}\right) = \frac{E_c V_0}{C} \tag{3.56}$$

As the pressure wave propagates upstream along the pipe at speed C, the water behind the wave front is immediately brought to a stop from the initial velocity of V_0. The total mass of water involved in this sudden change of speed from V_0 to zero in time Δt is $m = \rho \cdot AC \cdot \Delta t$. Applying Newton's second law to this mass, we have

$$\Delta P \cdot A = m\frac{\Delta V}{\Delta t} = \rho \cdot A \cdot C \cdot \Delta t \frac{(V_0 - 0)}{\Delta t} = \rho ACV_0$$

or

$$C = \frac{\Delta P}{\rho V_0}$$

Substituting the above value of C into Eq. (3.56), we have

$$\Delta P = E_c V \frac{\rho V_0}{\Delta P}$$

$$(\Delta P)^2 = \rho E_c V_0^2$$

or

$$H = \frac{\Delta P}{\rho g} = \frac{V_0}{g}\sqrt{\frac{E_c}{\rho}} = \frac{V_0}{g}C \tag{3.57}$$

which is the pressure head caused by the water hammer. The formula is applicable to rapid valve movement ($t \leq 2L/C$).

For closure time $t > 2L/C$, ΔP will not develop fully because the reflected negative wave arriving at the valve will compensate for the pressure rise. For such slow valve closures, the maximum water hammer pressure may be calculated by the Allievi formula.[*]

Applying the water hammer formula to pipe flow problems, we shall first determine the energy gradient line and the hydraulic gradient line for the pipe system under steady flow conditions, as shown in Figure 3.38.

Figure 3.38.   Water hammer pressure on a pipeline.

[*]L. Allievi, *The Theory of Water Hammer* (translated by E. E. Halmos), ASME, 1929. The maximum water hammer pressure calculated by the Allievi formula is

$$\Delta P = P_0\left(\frac{N}{2} + \sqrt{\frac{N^2}{4} + N}\right) \tag{3.58}$$

where P_0 is the static-state pressure in the pipe, and

$$N = \left(\frac{\rho L V_0}{P_0 t}\right)^2 \tag{3.59}$$

The total plate pressure experienced by the pipe is

$$P = \Delta P + P_0$$

The steady-state pressure head at any point X along the pipeline is represented by P_0/γ, which is the vertical distance between the point and the hydraulic gradient line; and the velocity head, $V_0^2/2g$, is the vertical distance between the energy gradient line and the hydraulic gradient line. When the peak water hammer pressure wave finally arrives at P, the water velocity behind the wave front becomes zero; hence, all the energy stored in this section of the pipe is in the form of pressure energy. The pressure head experienced by the pipe at point P is the total height of the water column PS. The following example illustrates this point.

Example 3.15

A steel pipe 1500 m long laid on a uniform slope has a 0.5-m diameter and a 5-cm wall thickness. The pipe carries water from a reservoir and discharges it into the air at an elevation 50 m below the reservoir free surface. A valve installed at the downstream end of the pipe allows a flow rate of 0.8 m³/s. If the valve is completely closed in 1.4 s, calculate the maximum water hammer pressure at the valve and at the midlength of the pipe. Neglect longitudinal stresses.

Solution

From Eq. (3.52a),

$$\frac{1}{E_c} = \frac{1}{E_b} + \frac{1}{E_p t}$$

where $E_b = 2.17 \cdot 10^9$ N/m², and $E_p = 1.9 \cdot 10^{11}$ N/m², as shown in Table 3.6. The above equation may thus be written as

$$\frac{1}{E_c} = \frac{1}{2.17 \cdot 10^9} + \frac{0.5}{(1.9 \cdot 10^{11}) \cdot 0.05}$$

Hence

$$E_c = 1.9 \cdot 10^9 \text{ N/m}^2$$

From Eq. (3.52) we have the speed of wave propagation along the pipe:

$$C = \sqrt{\frac{E_c}{\rho}} = \sqrt{\frac{1.9 \cdot 10^9}{1000}} = 1378 \text{ m/s}$$

The time required for the wave to return to the valve is t:

$$t = \frac{2L}{C} = \frac{2 \cdot 1500}{1378} = 2.17 \text{ s}$$

Since the water velocity in the pipe before valve closure is

$$V_0 = \frac{0.80}{\frac{\pi}{4} \cdot (0.5)^2} = 4.07 \text{ m/s}$$

the maximum water hammer pressure at the valve can be calculated.

$$\Delta P = \rho V_0 C = 1000 \cdot 4.07 \cdot 1378 = 5.61 \cdot 10^6 \text{ (N/m}^2)$$

Example 3.16

A cast iron pipe with 20 cm diameter and 15 mm wall thickness is carrying water when the outlet is suddenly closed. If the design discharge is 40 l/s, calculate the water hammer pressure rise for:

(a) rigid pipe wall;
(b) neglecting the longitudinal stress;
(c) pipeline to have expansion joints throughout the length.

Solution

$$A = \frac{\pi}{4} \cdot 0.2^2 = 0.0314$$

hence,

$$V = \frac{Q}{A} = \frac{0.04}{0.0314} = 1.274 \text{ m/s}$$

(a) For rigid pipe wall, $Dk/E_p t = 0$, Eq. (3.53) gives the following relation

$$\frac{1}{E_c} = \frac{1}{E_b}, \quad \text{or} \quad E_c = E_b = 1.6 \cdot 10^9 \text{ N/m}^2$$
$$\rho = 10^3 \text{ kg/m}^3$$

From Eq. (3.52), we can calculate the speed of pressure wave,

$$C = \sqrt{\frac{E_c}{\rho}} = \sqrt{\frac{1.6 \cdot 10^9}{10^3}} = 1265 \text{ m/s}$$

From Eq. (3.57), we can calculate the rise of water hammer pressure

$$H = \frac{VC}{g} = \frac{1.274 \cdot 1.265}{9.81} = 164 \text{ m (H}_2\text{O)}$$
$$\therefore \quad P = \gamma H = 1.61 \cdot 10^6 \text{ N/m}^2$$

(b) For pipes with no longitudinal stress, $K = 1$, we may use Eq. (3.53a),

$$E_c = \cfrac{1}{\left(\cfrac{1}{E_b} + \cfrac{D}{E_p t}\right)} = \cfrac{1}{\left(\cfrac{1}{1.6 \cdot 10^9} + \cfrac{0.2}{(1.6 \cdot 10^{11})(0.015)}\right)} = 1.41 \cdot 10^9$$

and

$$C = \sqrt{\frac{E_c}{\rho}} = 1188 \text{ m/s}$$

Hence, the rise of water hammer pressure can be calculated

$$H = \frac{VC}{g} = \frac{1.274 \cdot 1188}{9.81} = 154 \text{ m (H}_2\text{O)}$$

and

$$P = \gamma H = 1.51 \cdot 10^6 \text{ N/m}^2$$

(c) For pipes with expansion joints, $K = (1 - 0.5 \cdot 0.25) = 0.875$. From Eq. (3.53),

$$E_c = \cfrac{1}{\cfrac{1}{E_b} + \cfrac{0.875 D}{E_p t}} = \cfrac{1}{\left(\cfrac{1}{1.6 \cdot 10^9} + \cfrac{(0.875)(0.2)}{(1.6 \cdot 10^{11})(0.015)}\right)} = 1.43 \cdot 10^9$$

and

$$C = \sqrt{\frac{E_c}{\rho}} = 1197$$

Eq. (3.57) is used to calculate the rise of water hammer pressure

$$H = \frac{VC}{g} = \frac{1.247 \cdot 1197}{9.81} = 155 \text{ m (H}_2\text{O)}$$

Hence,

$$P = \gamma H = 1.53 \cdot 10^6 \text{ N/m}^2$$

In water hammer analysis the time history of pressure oscillation in the pipeline is determined. Because of the friction between the oscillatory water mass and the pipe wall, the pressure–time pattern is modified, and the oscillation gradually dies out. A typical pressure oscillation is shown in Figure 3.39.

In reality, a valve cannot be closed completely in zero time. The time required for closure of a valve is a certain period, t_c. The water hammer pressure increases gradually with the rate of closure of the valve. A typical valve closure curve is shown in Figure 3.40.

Figure 3.39.   Friction effect on pressure–time pattern.

Figure 3.40.   Typical valve closure curve.

If t_c is smaller than the time required for the wave front to make a round trip along the pipeline and return to the valve site ($t_c < 2L/C$), the operation is defined as *rapid closure*. The shock pressure will reach its maximum value. The computation of a quick closure operation is the same as that of an instantaneous closure. For slow closure operation, however, $t_c > 2L/C$, the pressure wave returns to the valve site before the closure is completed. A certain amount of water continuously passes through the valve when the pressure wave returns. As a result, the pressure wave pattern will be altered. A complete treatment of the water hammer phenomenon, with consideration of friction and slow valve closure operation, may be found elsewhere.*

PROBLEMS

3.12.1. A 500-m-long steel pipe carries water from a reservoir that has a water elevation at 55 m above the exit of the pipe. The pipe has a diameter of 40 cm and a wall thickness of 2 cm. A gate valve at the downstream end is slowly closed during a 30-s period. Determine the maximum water hammer pressure.

3.12.2. A horizontal pipe 30 cm in diameter and 420 m long has a wall thickness of

*Wylie, E. B. and V. L. Streeter, *Fluid Transients* (New York: McGraw-Hill, 1978).

of 1 cm. The pipe is steel and carries water from a reservoir to a level 100 m below to discharge into open air. A check valve of the lift type is installed at the downstream end. Determine the discharge when the valve is fully opened. Calculate the maximum pressure on the valve if it closes in a 0.5-s period.

3.12.3. Calculate the water hammer pressure and the speed of the pressure wave in Prob. 3.12.2. Assume that:
  (a) the pipe is free to move longitudinally;
  (b) the pipe is anchored against longitudinal motion;
  (c) the pipe has expansion joints.

PROJECT PROBLEM

From a reservoir 1200 m away a 15-cm diameter main pipe supplies water to six multistory buildings in an industrial park. The reservoir is 60 m above the datum elevation. The positions of the buildings are shown in Figure 3.41. The height and

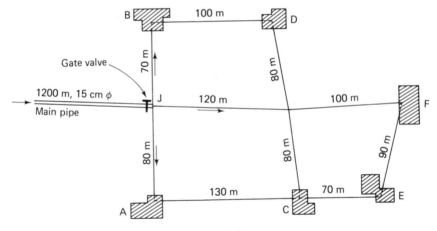

Figure 3.41

the water demand of each building are as follows:

Buildings	A	B	C	D	E	F
Height (m)	9.4	8.1	3.2	6.0	9.6	4.5
Water Demand (l/s)	5.0	6.0	3.5	8.8	8.0	10.0

If commercial steel pipes are used for the network (downstream from the junction J), determine the size of each pipeline. A gate valve ($K = 0.15$ when fully opened) is installed in the main pipe immediately upstream from the junction J. Determine the material to be used for the main pipe. Determine the water hammer pressure if the valve is suddenly closed. What must the minimum wall thickness of the pipe be in order to withstand the pressure?

4

WATER PUMPS

Water pumps are devices designed to convert mechanical energy to hydraulic energy. All forms of water pumps may be classified into two basic categories:

1. Turbo-hydraulic pumps.
2. Positive-displacement pumps.

Turbo-hydraulic pumps include mainly the centrifugal pumps, the propeller pumps, and the jet pumps. Analysis of the turbo-hydraulic machines is a problem involving hydraulic principles. Positive-displacement pumps move fluid strictly by precise machine displacements such as a gear system rotating with a closed housing (screw pumps) or a piston moving in a sealed cylinder (reciprocal pumps). Analysis of the positive-displacement pumps involves purely mechanical concepts and does not require the detailed knowledge of hydraulics. In this chapter only the first category, which constitutes most of the water pumps used in modern hydraulic engineering, will be treated in detail.

4.1 POWER AND EFFICIENCY OF HYDRAULIC PUMPS

The two basic equations for power of turbo-hydraulic pumps are (1) the power input to the pump:

$$P_i = \omega T \tag{4.1}$$

where ω is the angular velocity of the pump and T is the torque applied to the pump by a motor, and (2) the power output of the pump:

$$P_o = \gamma Q H \tag{4.2}$$

where γ is the unit weight of the liquid, Q is the volume flow rate of discharge, and H is the total energy head added to the liquid by the pump.

The *efficiency* of a pump is defined by the ratio of the output power to the input power of the pump.

$$e_p = \frac{P_o}{P_i} = \frac{\gamma Q H}{\omega T} \tag{4.3}$$

A hydraulic pump is usually driven by a motor. The efficiency of the motor is defined by the ratio of the power applied to the pump by the motor, P_i, to the power input to the motor, P_m.

$$e_m = \frac{P_i}{P_m} \tag{4.4}$$

The *overall efficiency* of the pump system is thus

$$e = e_p e_m = \left(\frac{P_o}{P_i}\right) \cdot \left(\frac{P_i}{P_m}\right) = \frac{P_o}{P_m} \tag{4.5}$$

or

$$P_o = e \cdot P_m \tag{4.5a}$$

The values of the e's are always less than unity due to friction and other energy losses that occur in the system.

Example 4.1

A centrifugal pump delivers 0.3 m³/s of water against an elevation head of 20 m when rotating at the angular velocity of 1500 rpm. If the torque measured at the driving shaft of the pump is 500 N-m, determine the efficiency of the pump.

Solution

The power transmitted into the fluid is the output power of the pump.

$$P_o = \gamma QH = (9.81 \text{ kN/m}^3)(0.3 \text{ m}^3/\text{s})(20 \text{ m}) = 58.86 \text{ kW}$$

The power input to the pump is

$$P_i = \omega T = \left(2\pi \cdot \frac{1500}{60}\right)(500 \text{ N-m}) = 78{,}500 \text{ W} = 78.5 \text{ kW}$$

The efficiency of the pump may be obtained from Eq. (4.3):

$$e_p = \frac{P_o}{P_i} = \frac{58.86}{78.5} \cdot 100\% = 75\%$$

PROBLEMS

4.1.1. A centrifugal pump delivers 2.5 m³/s against an elevation head of 20 m. If the pump operates at 85% efficiency, determine the input power at the pump shaft.

4.1.2. The pump in Prob. 4.1.1 is driven by a 750-kW electric motor. Determine the efficiency of the motor and the efficiency of the pump-motor system.

4.2 CENTRIFUGAL PUMPS

The fundamental principle of the centrifugal pump was first demonstrated by Demour in 1730. The simple pump, as shown in Figure 4.1, consisted of two straight pipes in the form of a tee. If the tee is primed with a liquid and the lower end of the tee is submerged, the rotation of the horizontal arms develops a centrifugal force which, when it overcomes the gravitational force on the liquid, moves the liquid upward through the vertical shaft and discharges it through the horizontal arms.

Modern centrifugal pumps consist basically of two parts:

1. The rotating element which is commonly called the *impeller.*

2. The *housing* that encloses the rotating element and seals the pressurized liquid inside.

Figure 4.1. Demour's centrifugal pump.

(a)

(b)

Figure 4.2. Cross sections of a centrifugal pump.

The power of a pump is supplied by a motor to the shaft of the impeller. The rotary motion of the impeller creates a centrifugal force that enables the liquid to enter the pump at the low-pressure region near the center (*eye*) of the impeller and to move along the direction of the impeller vanes toward the higher-pressure region near the outside of the housing surrounding the impeller, as shown in Figure 4.2(a). The housing is designed with a gradually expanding spiral shape so that the entering liquid is led toward the discharge pipe with minimum loss while the kinetic energy in the liquid is converted into pressure energy (Figure 4.2).

The theory of centrifugal pumps is based on the *principle of angular momentum conservation*. Physically, the term momentum (which usually refers to the linear momentum) is defined as the product of a mass and its velocity, or

$$\text{momentum} = (\text{mass})(\text{velocity})$$

The angular momentum with respect to a fixed axis of rotation may thus be defined as the moment of the linear momentum with respect to the axis:

$$\text{angular momentum} = (\text{momentum}) \cdot (\text{radius})$$
$$= (\text{mass})(\text{velocity}) \cdot (\text{radius})$$

The principle of conservation of angular momentum requires that the time rate of change of angular momentum in a body of fluid be equal to the torque resulting from the external force acting on the body. This relationship may be expressed as

$$\text{torque} = \frac{(\text{mass}) \cdot (\text{velocity}) \cdot (\text{radius})}{\text{time}} = \rho \left(\frac{\text{volume}}{\text{time}} \right) \cdot (\text{velocity}) \cdot (\text{radius})$$

The diagram in Figure 4.3 can be used to analyze this relationship.

The angular momentum (or moment of momentum) for a small fluid mass per unit time ($\rho \, dQ$) is

$$(\rho \, dQ)(V \cos \alpha)(r)$$

where $V \cos \alpha$ is the tangential component of the absolute velocity. For the total fluid mass that enters the pump in unit time, the angular momentum can be evaluated by the following integral:

$$\rho \int_{Q} rV \cos \alpha \, dQ$$

The torque applied to a pump impeller must equal the difference of angular momentum at the inlet to and outlet from the impeller. It may be expressed as follows:

$$T = \rho \int_{Q} r_o V_o \cos \alpha_o \, dQ - \rho \int_{Q} r_i V_i \cos \alpha_i \, dQ \tag{4.6}$$

For steady flow and uniform conditions around the impeller, $r_o V_o \cos \alpha_o \, dQ$ and $r_i V_i \cos \alpha_i \, dQ$ have constant values. Equation (4.6) may be simplified to

$$T = \rho Q (r_o V_o \cos \alpha_o - r_i V_i \cos \alpha_i) \tag{4.7}$$

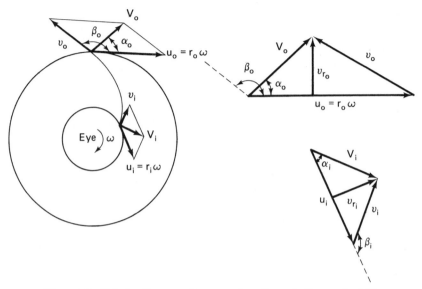

Figure 4.3. Velocity diagram of a pump impeller. u is the speed of the impeller vane ($u = r\omega$); v is the relative velocity of the liquid with respect to the vane; V is the absolute velocity of the liquid, a vector sum of u and v; β_o is the vane angle at the exit; β_i is the vane angle at the entrance; r is the radius measured from the center of the impeller; $r = r_i$ is the radius of the impeller eye at the entrance; and $r = r_o$ is the radius of the impeller at the exit.

Let ω be the angular velocity of the impeller. The power input to the pump can be computed by Eqs. (4.1) and (4.7).

$$P_i = \omega T = \rho Q \omega (r_o V_o \cos \alpha_o - r_i V_i \cos \alpha_i) \tag{4.8}$$

The output power of a pump is usually expressed in terms of the pump discharge and the total energy head that the pump imparts to the liquid, H. As previously discussed, the energy head in fluid can usually be expressed as the sum of the three forms of hydraulic energy: the kinetic energy, $V^2/2g$, the pressure energy, p/γ, and the elevation energy, h.

$$H = \frac{V^2}{2g} + \frac{p}{\gamma} + h$$

The *pump output power* may be computed by Eq. (4.2):

$$P_o = \gamma Q H = \gamma Q \left(\frac{V^2}{2g} + \frac{p}{\gamma} + h \right)$$

where V, P, and h, are, respectively, the velocity, pressure, and elevation added to the liquid by the pump.

The *polar vector diagram* (Figure 4.3) is generally used in analyzing the vane geometry and its relationship to the flow. As designated previously, the subscripts i and o are used, respectively, for the inlet and exit flow conditions. u represents the peripheral velocity of the impeller, v represents the water velocity relative to the vane blade (hence, in the direction of the blade), and V is the absolute water velocity. V_t is the tangential component of the absolute velocity, and V_r is the radial component. Theoretically, the energy loss at the inlet reaches its minimum value when water enters the impeller without whirl. This is achieved when the impeller is operated at such a speed that the absolute water velocity at the inlet is in the radial direction.

Example 4.2

A centrifugal pump has the following characteristics: $r_i = 12$ cm, $r_o = 40$ cm, $\beta_i = 118°$, $\beta_o = 140°$. The width of the impeller vanes is 10 cm and is uniform throughout. At the angular speed of 550 rpm the pump delivers 0.98 m³/s of water between two reservoirs with a 25-m elevation difference. Determine the efficiency of the pump at this stage of operation.

Solution

The peripheral speeds of the vanes at the entrance and at the exit of the impeller are, respectively,

$$u_i = \omega r_i = 2\pi \cdot \frac{550}{60 \text{ s}} \cdot 0.12 \text{ m} = 6.91 \text{ m/s}$$

$$u_o = \omega r_o = 2\pi \cdot \frac{550}{60 \text{ s}} \cdot 0.40 \text{ m} = 23.04 \text{ m/s}$$

and the radial velocity of the water may be obtained by applying the continuity equation: $Q = A_i V_{r_i} = A_o V_{r_o}$, where $A_i = 2\pi Br_i$ and $A_o = 2\pi Br_o$. It can be easily shown that $V_{r_i} = v_{r_i}$ and $V_{r_o} = v_{r_o}$.

$$v_{r_i} = \frac{Q}{A_i} = \frac{Q}{2\pi Br_i} = \frac{0.98}{2\pi \cdot 0.1 \cdot 0.12} = 13.00 \text{ m/s}$$

$$v_{r_o} = \frac{Q}{A} = \frac{Q}{2\pi Br_o} = \frac{0.98}{2\pi \cdot 0.1 \cdot 0.4} = 3.90 \text{ m/s}$$

From the velocity vector diagram (see Figure 4.3)

$$v_{t_i} = \frac{v_{r_i}}{\tan \beta_i} = \frac{13.00}{\tan 118°} = -6.91 \text{ m/s}$$

$$v_{t_o} - \frac{v_{r_o}}{\tan \beta_o} = \frac{3.90}{\tan 190°} = -4.65 \text{ m/s}$$

and

$$V_i = \sqrt{v_{r_i}^2 + (u_i + v_{t_i})^2} = \sqrt{(13.00)^2 + (0.00)^2} = 13.00 \text{ m/s}$$

$$\alpha_i = \tan^{-1} \frac{v_{r_i}}{(u_i + v_{t_i})} = \tan^{-1}\left(\frac{13.00}{0.00}\right) = 90°$$

$$\cos \alpha_i = 0$$

$$V_o = \sqrt{v_{r_o}^2 + (u_o + v_{t_o})^2} = \sqrt{(3.90)^2 + (18.39)^2} = 18.80 \text{ m/s}$$

$$\alpha_0 = \tan^{-1} \frac{v_{r_o}}{u_o + v_{t_o}} = \tan^{-1}\left(\frac{3.90}{18.93}\right) = 11.97°$$

$$\cos \alpha_o = 0.978$$

Applying Eq. (4.8), we get

$$P_i = \rho Q \omega (r_o V_o \cos \alpha_o - r_i V_i \cos \alpha_i)$$

$$P_i = 1000 \cdot 0.98 \cdot 2\pi \cdot \frac{550}{60} \cdot (0.40 \cdot 18.8 \cdot 0.978) = 415{,}120 \text{ W}$$

$$= 415.12 \text{ kW}$$

Applying Eq. (4.2) and assuming the only considerable head is the elevation, we get $H = h$:

$$P_o = \gamma Q H = 9.81 \text{ (kN/m}^3) \cdot 0.98 \text{ (m}^3/\text{s)} \cdot 25 \text{ (m)} = 240.35 \text{ kW}$$

From Eq. 4.3

$$e = \frac{P_o}{P_i} = \frac{240.35}{415.12} = 0.579 \cong 58\%$$

In Figure 4.2 the total energy head at the entrance to the pump is represented by

$$H_i = \frac{P_i}{\gamma} + \frac{V_i^2}{2g}$$

and the total energy head at the discharge is

$$H_d = h_d + \frac{P_d}{\gamma} + \frac{V_d^2}{2g}$$

The difference between the two is the amount of energy that the pump imparted to the liquid.

$$H = H_d - H_i = \left(h_d + \frac{P_d}{\gamma} + \frac{V_d^2}{2g}\right) - \left(\frac{P_i}{\gamma} + \frac{V_i^2}{2g}\right) \qquad (4.9)$$

The efficiency of a centrifugal pump depends largely on the particular design of the vane blades and the pump housing. It also depends on the conditions under which the pump operates. A typical pump characteristic chart is shown in Figure 4.4. It demonstrates the fact that each pump is designed for a specific range of discharge. Pump efficiency drops sharply when the discharge exceeds the design discharge range.

Figure 4.4. Typical centrifugal pump characteristic curves.

Example 4.3

A particular pump has the characteristics as shown in Figure 4.4. It is operated by a 100-kW electric motor with 65% efficiency. Determine the discharge, the dynamic head delivered, and the efficiency of the pump.

Solution

Applying Eq. (4.4), we get

$$P_i = e_m P_m = 0.65 \cdot 100 \text{ kW} = 65 \text{ kW}$$

Using the curves provided in Figure 4.4, we have

$$Q = 0.2 \text{ m}^3/\text{s} \qquad \text{(corresponding to 65 kW)}$$

In the same figure, for this discharge, the dynamic head and the efficiency are given. Respectively, they are

$$H = 19.6 \text{ m} \quad \text{and} \quad e_p = 40\%$$

PROBLEMS

4.2.1. A centrifugal pump with uniform impeller thickness 0.1 m, inlet radius $r_i = 30$ cm, outlet radius $r_o = 75$ cm, $\beta_i = 120°$, $\beta_o = 135°$ delivers a discharge of 2 m³/s against a head of 10 m. If the pump rotates at such a speed that no tangential velocity component exists at the inlet, what is the rotational speed? Calculate the efficiency of the pump.

4.2.2. At the outlet of a centrifugal pump impeller the radius is 60 cm and the width is 15 cm. If the velocity of water at the outlet (the absolute velocity) is measured to be 40 m/s at an angle of 60° measured with respect to the radial line, calculate the torque exerted on the impeller by the outlet flow.

4.2.3. A centrifugal pump impeller has an inlet diameter of 50 cm and outlet diameter of 150 cm. With $\beta_i = 135°$ and $\beta_o = 150°$, the pump is rotating at an angular velocity of 100 rad/s. The impeller has a uniform thickness of 30 cm. If the radial velocity component v_{r_i} is the same magnitude as the tangential velocity component v_{t_i}, calculate the discharge of the pump and the power input to the shaft of the pump.

4.2.4. If the efficiency of the pump in Prob. 4.2.3 is at 85%, calculate the head against which the discharge is delivered.

4.3 PROPELLER (AXIAL-FLOW) PUMPS

A rigorous mathematical analysis for designing propellers based strictly on the energy-momentum relationship is not available. However, the application of the basic *principle of impulse-momentum* provides a simple means of describing their operations.

Linear impulse is defined as the integral of the product of the force and the time, dt, from t' to t'' during which the force acts on the body.

$$I = \int_{t'}^{t''} F\, dt$$

If a constant force is involved during the time period, T, the impulse may be simplified to

$$(\text{impulse}) = (\text{force})(\text{time})$$

The principle of impulse-momentum requires that *the linear impulse of a force (or force system) acting on a body during a time interval be equal to the change in linear momentum in the body during that time.*

$$(\text{force})(\text{time}) = (\text{mass})(\text{velocity change})$$

or

$$(\text{force}) = \frac{(\text{mass})(\text{velocity change})}{(\text{time})} \tag{4.10}$$

The relationship may be applied to a body of fluid in steady motion by taking a volume between any two sections as shown in Figure 4.5. The factor, (mass)/(time), may be expressed as the mass involved per unit time, or,

$$\frac{(\text{mass})}{(\text{time})} = \frac{(\text{density})(\text{volume})}{(\text{time})} = (\text{density})(\text{discharge}) = \rho Q$$

and the change in velocity is thus the change of the fluid velocity between the two ends of the control volume.

$$(\text{velocity change}) = V_i - V_f$$

Substituting the above relationships into Eq. (4.10), we have

$$\Sigma F = (\text{force}) = \rho Q(V_i - V_f) \qquad (4.11)$$

Figure 4.5 shows schematically a propeller pump installed in a horizontal position. Four sections are selected along the pump tube to demonstrate the energy relationship. The system can be analyzed readily by the principle of impulse-momentum.

While the fluid moves from section 1 to section 2, the velocity increases and the pressure drops according to the Bernoulli principle of energy balance:

$$\frac{P_1}{\gamma} + \frac{V_1^2}{2g} = \frac{P_2}{\gamma} + \frac{V_2^2}{2g}$$

Between the two adjacent sections 2 and 3, energy is added to the fluid by the propeller. The energy is added to the fluid in the form of pressure head, which results in a higher pressure immediately downstream from the propeller. Further downstream at the exit end of the pump (section 4), the flow

Figure 4.5. Propeller pump.

condition is more stable, and a slight drop in pressure head may result from both the head loss between sections 3 and 4 and a slight increase in mean stream velocity.

Applying the impulse-momentum relationship, Eq. (4.11), between sections 1 and 4, we may write the following equation:

$$P_1 A_1 + F - P_4 A_4 = \rho Q (V_4 - V_1) \tag{4.12}$$

where F is the force exerted to the fluid by the propeller. The right-hand side of Eq. (4.12) drops out when the pump is installed in a conduit of uniform diameter, for which $V_1 = V_4$ and

$$F = (P_1 - P_4)A$$

In this case, the force imparted by the pump is used totally to generate pressure.

$$\frac{P_1}{\gamma} + \frac{V_1^2}{2g} = \frac{P_2}{\gamma} + \frac{V_2^2}{2g} \tag{4.13}$$

and between sections 3 and 4, we may write

$$\frac{P_3}{\gamma} + \frac{V_3^2}{2g} = \frac{P_4}{\gamma} + \frac{V_4^2}{2g} \tag{4.14}$$

Subtracting Eq. (4.13) from Eq. (4.14), and noting that $V_2 = V_3$ for the same cross-section area, we have

$$\frac{P_3 - P_2}{\gamma} = \left(\frac{P_4}{\gamma} + \frac{V_4^2}{2g}\right) - \left(\frac{P_1}{\gamma} + \frac{V_1^2}{2g}\right) = H \tag{4.15}$$

where H is the total energy imparted to the fluid by the pump. The total power output from the pump is, therefore,

$$P_o = \gamma Q H = Q(P_3 - P_2) \tag{4.16}$$

The efficiency of the pump may be computed by the ratio of the output power of the pump to the input power by the motor.

Propeller pumps are usually used for low head (under 12 m), high-capacity (above 20 l/s) applications. Figure 4.9 shows the relative regions of application for different types of pumps.

More than one set of propeller blades may be mounted on the same axis of rotation in a common housing to form a *multistage propeller pump*, as shown in Figure 4.6. These pumps are capable of delivering a large quantity of water over a great elevation difference. The pumps are usually designed for

Figure 4.6. Multistage propeller pump.

self-priming operations and are used most frequently for pumping deep water wells.

Example 4.4

A 3-m diameter propeller pump is installed to deliver a large quantity of water over a 2.6-m elevation head. If the shaft power supplied to the pump is 1500 kW and if the pump operates at 80% efficiency, determine the discharge rate.

Solution

The energy imparted to the flow by the pump is given by Eq. 4.2:

$$P_o = e_p P_i = 0.8 \cdot 1500 \text{ kW} = 1200 \text{ kW} = 1,200,000 \text{ W}$$

Applying Eq. 4.14, we find

$$P_o = \gamma Q H = \gamma Q \left(h + \Sigma k \frac{V^2}{2g} \right)$$

For $k_{in} = 0.5$ and $k_{out} = 1.0$, we have

$$P_o = \gamma Q\left(h + 1.5\frac{Q^2}{2gA^2}\right); \qquad \gamma = 9810 \text{ N/m}^3$$

and

$$1,200,000 = 9810Q\left(2.6 + 1.5\frac{Q^2}{(1.1\pi)^2 \cdot 2 \cdot 9.81}\right)$$

By trial, the above equation may be solved to obtain

$$Q = 30.45 \text{ m}^3/\text{s}$$

4.4 JET MIXED-FLOW PUMPS

The jet pump utilizes a high-pressure stream. The pressurized fluid ejects from a nozzle at high speed into the surrounding fluid, transferring its energy to the surrounding fluid and bringing it to delivery. Jet pumps are usually used in combination with a centrifugal pump, which supplies the high-pressure stream, and can be used to lift liquid in deep wells. Jet pumps are usually compact in size and light in weight and are sometimes used in construction work for dewatering the field. Since the energy loss during the mixing procedure is heavy, the efficiency of the jet pump is normally very low (rarely more than 25%). A schematic diagram of a jet pump is shown in Figure 4.7.

Low pressure and suction

Low pressure delivery

High pressure inlet

Figure 4.7. Schematic of a jet pump.

4.5 CAVITATION IN WATER PUMPS

One of the important considerations in pump installation design is the relative elevation between the pump and the water surface in the supply reservoir. The water in the suction line in a pumping installation is usually under pressure lower than atmospheric. (The phenomenon of cavitation becomes a potential danger whenever the water pressure at any location in the pumping system drops substantially below atmospheric pressure.) A common site of cavitation is near the tips of the impeller vanes where the velocity is very high. In regions of high velocities much of the pressure energy is converted to kinetic energy. This is added to the elevation difference between the pump and the supply reservoir and to the inevitable energy loss in the pipeline between the reservoir and the pump. The sum of those three items constitutes the *total suction head*, H_s, in a pumping installation, as shown in Figure 4.8. The total suction head may be expressed as

$$H_s = \frac{V_t^2}{2g} + H_p + \Sigma H_L \qquad (4.17)$$

where H_p is the elevation difference between the pump and the supply reservoir, ΣH_L is the total energy loss in the suction line, and V_t is the local velocity near the tip of the impeller vanes.

The value of H_s must be kept within a limit so that the pressure at any location in the pump is always above the vapor pressure of water; otherwise, the water will be vaporized and cavitation well occur. The vaporized water forms small vapor bubbles in the flow. These bubbles collapse when they reach the region of higher pressure. Violent vibrations may result from the collapse of vapor bubbles in water. Successive bubble breakup with considerable impact force may cause high local stresses on the metal surface of

Figure 4.8.   Total suction head on a pump.

the vane blades and the housing. These stresses cause surface pitting and will rapidly damage the pump.

To prevent cavitation, the pump should be installed at an elevation such that the total suction head is less than the difference between the atmospheric head and the water vapor pressure head, or

$$\left(\frac{p_{atm}}{\gamma} - \frac{p_{vapor}}{\gamma}\right) - H_s > 0$$

The maximum velocity near the tip of the impeller vanes is not usually assessed by the users. Pump manufacturers usually provide a value known commercially as the *net positive suction head* (NPSH), or H'_s, as shown in Figure 4.8. With the value of NPSH, the maximum pump elevation above the supply reservoir can be easily determined:

$$H_p = \frac{p_{atm}}{\gamma} - \left(\frac{p_{vapor}}{\gamma} + H'_s + \frac{V^2}{2g} + H_L\right) \tag{4.18}$$

The other parameter commonly used for preventing cavitation in a pump is described by the *cavitation parameter, σ*, which is defined as:

$$\sigma = \frac{\left(\frac{p_{atm}}{\gamma} - \frac{p_{vapor}}{\gamma}\right) - H_s}{H} \tag{4.19}$$

where H is the total dynamic head developed by the pump and the numerator is the net positive suction head (NPSH). The term H_s is similar to that expressed in Eq. (4.17), except that the velocity head here is the water velocity at the intake of the pump. The increase in velocity about the impeller vanes is already accounted for in the parameter σ. The value of σ for each type of pump is usually furnished by the manufacturer and is based on pump test data.

Substituting the relationship of Eq. (4.17) into Eq. (4.19), we may write

$$\sigma H = \frac{p_{atm}}{\gamma} - \frac{p_{vapor}}{\gamma} - \left(\frac{V_i^2}{2g} + H_p + \Sigma H_L\right) \tag{4.20}$$

where V_i is the highest speed of water at the entrance to the impeller. Rearranging Eq. (4.20), we get

$$H_p = \frac{p_{atm}}{\gamma} - \frac{p_{vapor}}{\gamma} - \frac{V_i^2}{2g} - \Sigma H_L - \sigma H \tag{4.21}$$

which will give the maximum allowable elevation of the intake (entrance to the impeller) above the surface of the supply reservoir. If the value determined by Eq. (4.21) is negative, then the pump must be placed at an elevation below the water surface elevation in the supply reservoir.

Example 4.5

A pump is installed at sea level in a 15-cm pipeline 300 m long to pump 0.060 m^3/s water at 20°C. The height is 25 m. If the pump has a 18-cm impeller intake diameter, has a cavitation parameter of $\sigma = 0.12$, and experiences a total head loss of 1.3 m at the suction side, determine the relative elevation between the pump intake and the water surface elevation in the supply tank. Assume the pipeline has $C_{HW} = 120$.

Solution

The head loss in the pipeline can be determined by the chart provided in Figure 3.9 (Hazen–Williams formula). For $Q = 0.06$ m^3/s, and $C_{HW} = 120$; $D = 0.15$ m and $L = 300$ m, we found the slope of the energy gradient line, $S = 0.086$ and $H_f = L \cdot S = 25.8$. The only local loss is at the exit, for which we may assume $k_e = 1.0$, as discussed in Chapter 3.

The velocities in the intake pipe and in the conduction pipe line are, respectively,

$$V_i = \frac{Q}{A_i} = \frac{0.06}{\pi(0.09)^2} = 2.36 \text{ m/s}$$

$$V_p = \frac{Q}{A_p} = \frac{0.06}{\pi(0.075)^2} = 3.40 \text{ m/s}$$

The total dynamic head developed by the pump can be determined by applying Bernoulli's equation to the reservoirs at both ends of the system:

$$\frac{V_1^2}{2g} + \frac{P_1}{\gamma} + h_1 + H = \frac{V_2^2}{2g} + \frac{P_2}{\gamma} + h_2 + \Sigma H_L$$

where subscripts 1 and 2 refer to the reservoirs at the supply end and delivery end, respectively. At the surface level of the reservoirs, we may write: $V_1 \simeq V_2 \simeq 0$ and $P_1 = P_2 = P_{atm}$; so

$$H = (h_2 - h_1) + \Sigma H_L = 25 + \left(1.3 + 25.8 + 1 \cdot \frac{3.40^2}{2g}\right) \simeq 52.7 \text{ m}$$

The vapor pressure is found in Table 1.1.

$$P_{vapor} = 0.023042 \text{ bar} = 2335 \text{ N/m}^2,$$

while

$$P_{atm} = 1 \text{ bar} = 101,357 \text{ N/m}^2$$

Finally, applying Eq. (4.21), we obtain the maximum relative elevation of the pump:

$$H_p = \frac{P_{atm}}{\gamma} - \frac{P_{vapor}}{\gamma} - \frac{V_i^2}{2g} - \Sigma H_{L_i} - \sigma H$$

$$= \frac{101,357}{9810} - \frac{2335}{9810} - \frac{(2.36)^2}{2 \cdot 9.81} - 1.3 - 0.12 \cdot 52.7 = 2.19 \text{ m}$$

PROBLEMS

4.5.1. Factory tests indicated that a cavitation parameter for a particular pump is $\sigma = 0.075$. The pump is installed to pump 60°C water at sea level. If the total head loss between the inlet and the suction side of the pump is 0.5 m, determine the allowable level of the pump intake relative to the supply reservoir water surface elevation. The discharge is 0.04 m³/s and the pipe system is the one used in Example 4.5.

4.5.2. The efficiency of a pump will drop suddenly if cavitation takes place in the pump. If this phenomenon is observed in a particular pump (with $\sigma = 0.08$) operating at sea level and the pump delivers 0.42 m³/s of water at 40°C, determine the sum of the gage pressure and the velocity head at the inlet. The dynamic head is 85 m and the suction pipe diameter is 30 cm.

4.6   SELECTION OF A PUMP

The efficiency of a pump depends on the discharge, head, and the power requirement of the pump. The ranges of application of each type of pump are shown in Figure 4.9.

The total head against which the pump delivers its discharge includes the elevation head and the head losses incurred in the system. The friction loss and other minor losses in the pipeline depend on the velocity of water in the pipe (Chapter 3), and the total head loss can be related to the discharge rate. The total head and discharge in a system may be plotted. For a given pipeline system (including the pump), a unique *H–Q* curve can be plotted, as shown in Example 4.6.

In selecting a particular pump for a given system, the design conditions are specified and a pump is selected for the range of applications. The *H–Q* curve is then matched to the pump performance chart (for example, Fig. 4.10) provided by the manufacturer. The matching point, *M*, indicates the actual working conditions. The selection process is demonstrated in the following example.

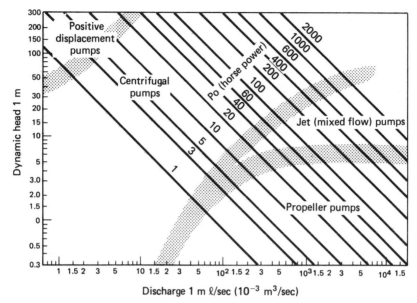

Figure 4.9. Heads, discharges, and power requirements of different types of pump.

Example 4.6

A pump will be used to deliver the discharge of 7 l/s of water between two reservoirs 1000 m apart with an elevation difference of 35 m. If commercial steel pipes of 10-cm diameter are used for the project, select the proper pump and determine the working condition for the pump based on the pump characteristics curves (H vs Q, Figure 4.10) and the pump performance chart (Figure 4.11), both provided by a manufacturer.

Solution

For commercial steel pipe, the roughness height, $e = 0.045$ mm (Figure 3.7). The flow velocity in the pipe is

$$V = \frac{Q}{A} = \frac{0.007 \text{ m}^3/\text{s}}{\left(\frac{\pi}{4} \cdot 0.1^2\right) \text{m}^2} = 0.89 \text{ m/s}$$

and the corresponding Reynolds number is

$$N_R = \frac{VD}{\nu} = \frac{0.89 \text{ m/s} \cdot 0.1 \text{ m}}{1 \cdot 10^6 \text{ m}^2/\text{s}} = 89{,}127 \simeq 9 \cdot 10^4$$

and

$$e/D = 0.045 \text{ mm}/100 \text{ mm} = 4.5 \cdot 10^{-4} = 0.00045.$$

The friction coefficient can be obtained from the Moody diagram (Figure 3.8).

$$f = 0.033$$

The pipe friction losses are, then,

$$h_f = f\left(\frac{L}{D}\right)\frac{V^2}{2g} = 0.033 \cdot \frac{1000}{0.1} \cdot \frac{0.89^2}{2 \cdot 9.81} = 13.32 \text{ m}$$

The total head that the pump must work against is

$$H = \text{(elevation difference)} + \text{(friction loss)}$$
$$= 35 + 13.32 = 48.32 \text{ m}$$

From the pump selection chart provided by the manufacturer (see, for example, Figure 4.11), pumps IV and V may be used for the project. The following quantities should be determined before making a selection:

Q (l/s)	V (m/s)	N_R	f	H_f	H
5	0.637	$6.9 \cdot 10^4$	0.036	7.95	42.5
6	0.764	$7.6 \cdot 10^4$	0.034	10.11	45.1
8	1.019	$10.2 \cdot 10^4$	0.032	16.92	51.9

These values are plotted, with H vs. Q, as a curve. Matching the curve with the characteristic curves of pumps IV and V, as provided by the manufacturer (Figure 4.10), we find that the possibilities are as follows:

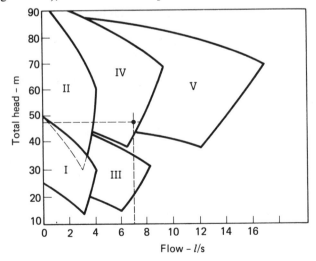

Figure 4.10. Pump selection chart.

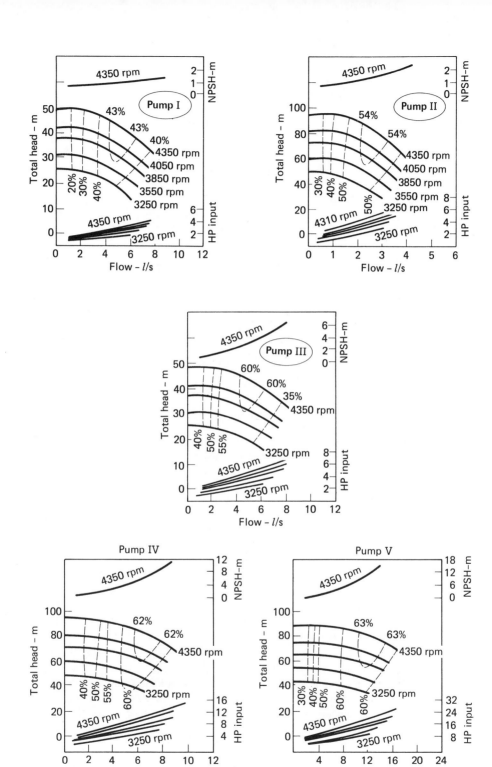

Figure 4.11. Characteristic curves of several pumps.

Pump IV at 3550 rpm:

$$Q = 6.8 \ l/s, \ H = 46 \ m;$$

hence,

$$P_i = 7.5 \ \text{hp and efficiency} = 60\%$$

Pump IV at 3850 rpm:

$$Q = 8.1 \ l/s, \ H = 52 \ m;$$

hence,

$$P_i = 10 \ \text{hp and efficiency} = 57\%$$

Pump V at 3250 rpm:

$$Q = 6 \ l/s, \ H = 45 \ m;$$

hence,

$$P_i = 7.5 \ \text{hp and efficiency} = 55\%$$

Pump V at 3550 rpm:

$$Q = 8.4 \ l/s, \ H = 54 \ m;$$

hence,

$$P_i = 10.5 \ \text{hp and efficiency} = 61\%$$

If a deviation of 3% is allowed for the design flow rate, the first case, (pump IV at 3550 rpm) is the choice since it best fits the conditions. Otherwise, we may use the second case (pump IV at 3850 rpm).

PROBLEMS

4.6.1. Select a pump to be used to supply water to a reservoir at a flow rate of 3 *l/s*. The elevation head is 20 m and the distance between the supply and delivery is 100 m. A ball check valve is used with 4-cm diameter galvanized iron pipeline. Determine the working conditions.

4.6.2. Select a pump and determine its working conditions for a job that demands a minimum discharge of 8 *l/s* against an elevation head of 50 m. The distance between the supply and delivery is 150 m. A ball check valve will be used in the system that consists of commercial steel pipes. Also select the most economical diameter for the pipeline if the total cost can be expressed as

$$C = d^{1.75} + 0.75P + 18$$

where *d* is the pipe diameter in cm and *P* is the power in horsepower.

4.7 PUMPS IN PARALLEL OR IN SERIES

As discussed in Section 4.2, the efficiency of a pump varies with the discharge rate of the pump and the height over which the delivery is made. The optimum efficiency of a pump can be obtained only over a limited range of operations (see Figure 4.4). To install a pumping station that can be effectively operated over a large range of fluctuations in both discharge and pressure, it may be advantageous to install several identical pumps at the station.

When several pumps are connected in parallel in a pipeline, the discharge is increased but the pressure head remains the same as with a single pump. It should be noted that two identical pumps operating in parallel may not double the discharge in a pipeline since the total head loss in a pipeline is proportional to the second power of discharge. The additional resistance in the pipeline will cause a reduction in the total discharge. Figure 4.12 shows schematically the operation of two identical pumps in parallel. The joint discharge of the two pumps is always less than twice the discharge of a single pump.

Pumps connected in series in a pipeline will increase the total output pressure, but the discharge will remain approximately the same as that of a single pump. A typical performance curve for two pumps connected in series is shown by curve C in Figure 4.12.

The efficiency of two (or more) pumps operating in parallel or in series is almost the same as that of the single pump. The installation can be arranged with one separate motor for each pump or with one motor to operate two

Figure 4.12. Typical performance curves of two pumps connected in parallel (B) and in series (C).

137

(or more) pumps. Multipump installations could be designed to perform either in-series or in-parallel operations with the same set of pumps. Figure 4.13 is a typical schematic of such an installation. For series operations, valve *A* is opened and valves *B* and *C* are closed; for parallel operations, valve *A* is closed and valves *B* and *C* are open.

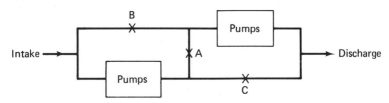

Figure 4.13. Schematic installation for operation either in series or in parallel.

Example 4.7

Two reservoirs are connected by a long pipeline in such a way that the total head between the two reservoirs can be expressed as $H = 20 + 6000Q^2$, where Q is the discharge in cubic meters per second and H is the total head in meters. The characteristic curve of the particular pump that can be used is given as follows:

Head (m)	70	60	50	40	30	20
Discharge (m³/s)	0	0.021	0.034	0.043	0.049	0.053
Efficiency (%)	0	69	80	68	47	30

Determine the discharge, head, and efficiency if we (1) use one pump, (2) use two pumps in series, and (3) use two pumps in parallel.

Solution

First, we plot the characteristic curves as shown in Figure 4.14. Then we plot the system curve in the same diagram. The intersection points define the working conditions.

(1) One pump:
   $Q = 0.047$ m³/s
   $H = 33$ m
   $e_p = 55\%$

(2) Two pumps in series:
   $Q = 0.054$ m³/s
   $H = 37$ m
   $e_p = 29\%$

(3) Two pumps in parallel:

$Q = 0.069$ m³/s   (0.0345 m³/s for each pump)

$H = 50$ m

$e_p = 78\%$

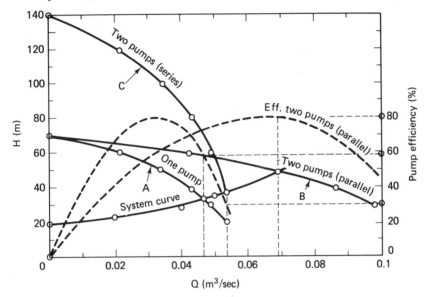

Figure 4.14

PROBLEMS

4.7.1. Two identical pumps have the characteristic curves shown in Figure 4.14. The pumps are connected in series and deliver water through a horizontal, 15 cm-diameter, 1000-m-long steel pipe into a reservoir in which the water level is 25 m above the pump. Neglect the minor losses in the system and determine the discharge.

4.7.2. Determine the discharge when the two pumps in Prob. 4.7.1 are connected in parallel.

4.8   SPECIFIC SPEED AND PUMP SIMILARITY

For a particular service, the selection of a pump is based on the required discharge rate and the head against which the discharge is delivered. To lift a large quantity of water over a relatively small elevation (for example, removing water from an irrigation canal onto a farm field) requires a high-capacity, low-stage pump. To pump a relatively small quantity of water against great heights (such as supplying water to a high-rise building) requires a low-capacity, high-stage pump. The designs of these two pumps are very different.

Generally speaking, impellers of relatively large radius and narrow flow passages transfer more kinetic energy head to the pressure head than impellers of smaller radius of large flow passages. Pumps designed with geometry allowing the water to exit the impeller in a radial direction impart more centrifugal acceleration to the flow than those that allow water to exit axially or at an angle. Thus, the relative geometry of the impeller and the pump housing determines the performance and the field application of a specific centrifugal pump.

Dynamic analysis (Chapter 8) shows that centrifugal pumps built with identical proportions but different sizes have similar dynamic performance characteristics which are consolidated into one number called a *shape number*. The shape number of a particular pump design is a dimensionless number defined as

$$S = \frac{\omega \sqrt{Q}}{(gH)^{3/4}} \tag{4.22}$$

where ω is the angular velocity of the impeller in radians per second, Q is the discharge of the pump in cubic meters per second, g is the gravitational acceleration in meters per second squared, and H is the total dynamic head in meters that the pump develops.

In engineering practice, however, the dimensionless shape number is not commonly used. Instead, most of the commercial pumps are specified by the term *specific speed*. The specific speed of a specific design of centrifugal pump can be defined in two different ways. Some manufacturers define the specific speed of a specific pump design as the speed of some unit of the design series of such a size that it delivers unit discharge at unit head. This way, the specific speed may be expressed as

$$\omega_s = \frac{\omega \sqrt{Q}}{H^{3/4}} \tag{4.23}$$

Other manufacturers define the specific speed of a specific design as the speed of a unit of the series of such a size that it produces unit power with unit head. This way, the speed is expressed as

$$\omega_s = \frac{\omega \sqrt{P_i}}{H^{5/4}} \tag{4.24}$$

Most of the commercial pumps manufactured in the United States are presently specified with the English units [gallons per minute (gpm), brake horsepower (bhp), feet (ft), and revolutions per minute (rpm)]. In metric units, cubic meters per second, kilowatts, meters, and revolutions per minute are usually used in the computations.

Normally, the specific speed is defined, one way or the other, at the optimum point of operation efficiency. In practice, pumps with high specific
speeds are generally used for large discharges at low-pressure heads, while
pumps with low specific speeds are used to deliver small discharge at high-
pressure heads. Centrifugal pumps with identical geometric proportions but
different sizes have the same specific speed.

Example 4.8

A centrifugal water pump operates at its optimum efficiency and delivers
2.5 m³/s over a height of 20 m. The pump has a 36-cm impeller diameter and
rotates at 3000 rpm. Compute the specific speed of the pump (1) in terms of
discharge and (2) in terms of power if the maximum efficiency of the pump
is 80%.

Solution

The given conditions are $Q = 2.5$ m³/s, $H = 20$, and $N = 3000$ rpm.
(1) Applying Eq. (4.23), we get

$$\omega_s = \frac{3000\sqrt{2.5}}{(20)^{3/4}} = 500 \text{ rpm}$$

(2) At 80% efficiency the shaft power is

$$P_i = \frac{\gamma QH}{0.80} = \frac{9810 \cdot 2.5 \cdot 20}{0.80} = 613{,}125 \text{ W} = 613 \text{ kW}$$

Applying Eq. (4.24), we get

$$\omega_s = \frac{3000\sqrt{613}}{(20)^{5/4}} \simeq 1760 \text{ rpm}$$

PROBLEMS

4.8.1. An homologous pump of the same design as in Example 4.8 has an impeller
diameter of 72 cm and has the same efficiency when operated at 1800 rpm.
Determine the discharge and shaft power required to operate the pump
against a head of 30 m.

4.8.2. Two centrifugal pumps of the same design are operated at the same efficiency.
Pump A is one-fourth the size of pump B in the corresponding dimensions.
When operating at 450 rpm, pump B delivers 2.4 m³/s against a 22-m head.
With the rotational speed of 1800 rpm what will be the discharge of pump A
when it is delivering water against the same head?

4.8.3. The design of a centrifugal water pump is studied by a 1/10 scale model in a hydraulic laboratory. At the optimum efficiency of 89% the model delivers 75.3 *l*/s of water against a 10-m head at 4500 rpm. If the prototype pump has a 72% efficiency at the rotational speed of 2250 rpm, determine the discharge and shaft power required to operate the pump under this condition.

PROJECT PROBLEM

A pumping system is designed to pump water from a 6-m-deep supply reservoir to a water tower 40 m above the ground. The system consists of a pump (or a combination of pumps), a 20-m-long pipeline with one elbow on the suction side of the pump, a 60-m-long pipeline, a gate valve, a check valve, and two elbows on the deliver side of the pump. The system is designed to pump 20 *l*/s of water, operating 350 days a year. Select a pump (or pumps) based on the characteristics provided in Figures 4.10 and 4.11, and a pipe size for optimum economy. Hazen-Williams coefficient 100 can be assumed for all pipe sizes listed below, and all elbows are 90 $(R/D = 2.0)$. Power cost is $0.04/kw-Hr. Motor efficiency is 85% for all sizes listed.

Pump	Cost	Motor	Cost
I	$700	6 HP	$200
II	900	8 HP	250
III	800	8 HP	200
IV	1020	15 HP	340
V	1240	30 HP	520

	20 cm	25 cm	30 cm
Pipe (10 m)	$120	$150	$180
Elbow	15	25	35
Gate valve	60	90	120
Check valve	80	105	130

5

WATER FLOW IN
OPEN CHANNELS

Open channel flow differs from pipe flow in one important aspect: Pipe flow usually fills the entire conduit and open channel flow must have a free surface. Normally, a free water surface is subject to the atmospheric pressure, which remains approximately a constant value throughout the entire length of the channel. Pipe flow, while confined in a closed conduit, experiences only the hydraulic pressure, which varies from one section to another along the pipeline.

In Figure 5.1 an open channel flow is schematically compared to a pipe flow. Figure 5.1(a) shows a pipe flow with two open-ended vertical tubes (piezometric tubes) installed through the pipe wall at an upstream section (1) and a downstream section (2). The water level in each tube represents the pressure head, P/γ, in the pipe at the section. The difference in water surface elevations in the two tubes indicates the *hydraulic gradient line* over these sections. The velocity head at each section is represented in the familiar form, $V^2/2g$, where V is the mean velocity, $V = Q/A$, at the section. The total energy head at any section is equal to the sum of the elevation (potential) head, h, the pressure head, P/γ, and the velocity head, $V^2/2g$. The difference in total energy head between the two sections is indicated as the *energy gradient line*. The amount of energy loss when water flows from section 1 to section 2 is indicated by h_L.

Figure 5.1(b) shows an open channel flow. The free water surface is subjected to only atmospheric pressure, which is commonly referred to as the

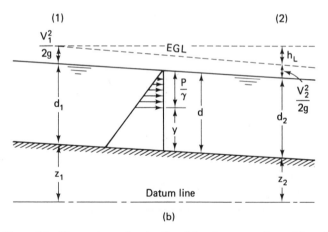

Figure 5.1. Comparison of a pipe flow (a) and an open channel flow (b).

zero pressure reference in hydraulic engineering practice. The pressure distribution at any section is directly proportional to the depth measured from the free water surface. In this case, the water surface line corresponds to the hydraulic gradient line in pipe flow.

To solve problems of open channel flow, we must seek the interdependent relationships between the slope of the channel bottom, the discharge, the water depth, and other given channel conditions.

5.1 CLASSIFICATION OF OPEN CHANNEL FLOWS

Open channel flows may be classified by either the time criterion or the space criterion.

Based strictly on the time criterion, open channel flows may be classified into two categories: *steady flows* and *unsteady flows*. In a steady flow the discharge and water depth at any section in the reach do not change with time during the period of interest. In unsteady flow the discharge and the water depth at any section in the reach change with time.

Based on the space criterion, an open channel flow is said to be *uniform flow* if the discharge and the water depth remain the same in every section in the channel reach. Uniform flows in open channel are mostly steady (Figure 5.2a); unsteady uniform flows (Figure 5.2b) are very rare in nature. A *varied flow* in open channel is one in which the water depth and/or the discharge change along the length of the channel. Varied open channel flow may be

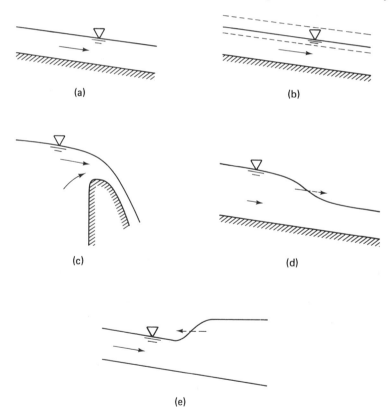

Figure 5.2. Classification of open channel flows. (a) Uniform flow. (b) Unsteady uniform flow. (c) Steady flow over a spillway. (d) Flood wave. (e) Tidal surge (bore).

either steady or unsteady. Steady flow over a spillway crest (Figure 5.2c) is a *varied steady flow*; a flood wave (Figure 5.2d) and a tidal surge (Figure 5.2e) are examples of *varied unsteady flows*.

PROBLEM

5.1.1. Using the time and space criterion, classify the following open channel flows during:

(a) The opening of a gate to let water into an otherwise uniform flow channel.

(b) Uniform rainfall over a sloped, long roof top.

(c) Flow in the gutter resulting from (b).

(d) Flow in the above gutter when the rainfall intensity increases with time.

5.2   UNIFORM FLOW IN OPEN CHANNELS

Uniform flow in an open channel must satisfy the following main features: (1) The water depth, water area, discharge, and the velocity distribution at all sections throughout the entire channel reach must remain unchanged and (2) the energy gradient line, the water surface line, and the channel bottom line must be parallel to each other. The slopes of these parallel lines are the same:

$$S_f = S_{w.s.} = S_0 = S$$

as shown in Figure 5.3.

Water in an open channel can reach the state of uniform flow only if no acceleration (or deceleration) takes place between sections. This is possible

Figure 5.3.   Force components in uniform open channel flow.

when the gravity force component and the resistance to the flow are equal and opposite in direction in the reach.

In two arbitrary sections, 1 and 2, in a uniform open channel, a *free-body diagram* can be taken between the two sections to show the balance of the force components of gravity and resistance (Figure 5.3).

The forces acting upon the free body in the direction of the flow include: (1) the hydrostatic pressure forces, F_1 and F_2, acting at each end of the free body; (2) the weight of the water body in the reach, W, which has a component, $W \sin \theta$, in the direction of the flow; and (3) the resistance force, F_f, exerted by the channel bed to the flow. A summation of all these force components gives the following equation:

$$F_1 + W \sin \theta - F_2 - F_f = 0 \qquad (5.1)$$

This equation can be further simplified, since there is no change in water depth in uniform flow between any two sections, 1 and 2. The hydrostatic forces at the two ends must be equal, $F_1 = F_2$. The total weight of the water body is

$$W = \gamma AL$$

where γ is the unit weight of water, A is the water cross-sectional area normal to the flow, and L is the length of the reach. In most open channels the channel slopes are small and the approximation, $\sin \theta = S$, is made. The slope S is also the energy gradient line slope. The gravity force component may thus be expressed as

$$W \sin \theta = \gamma ALS \qquad (5.1a)$$

The resistance force exerted by the channel boundaries may be expressed in terms of resisting force per unit area of the channel bed times the total channel bed area wetted by the flowing water. The wetted bed area is the product of the wetted perimeter, P, and the length of the channel reach, L.

In 1769 a French engineer, Antoine Chezy, assumed that the resisting force per unit area of the channel bed is proportional to the square of the mean velocity, KV^2, where K is a constant of proportionality. The total resistance force may thus be written as

$$F_f = \tau_0 PL = KV^2 PL \qquad (5.1b)$$

where τ_0 is the resisting force per unit area of the channel bed, also known as the *wall shear stress*.

Substituting Eq. (5.1a) and Eq. (5.1b) into Eq. (5.1), we have

$$\gamma ALS = KV^2 PL$$

or

$$V = \sqrt{\left(\frac{\gamma}{K}\right)\left(\frac{A}{P}\right)S}$$

In this equation, $A/P = R_H$, and $\sqrt{(\gamma/K)}$ may be represented by a constant, C. The above equation may thus be simplified to

$$V = C\sqrt{R_H S} \qquad (5.2)$$

in which R_H is the *hydraulic radius* of the channel cross section. The hydraulic radius is defined as the water area divided by the wetted perimeter for all shapes of open channel cross sections.

Equation 5.2 is the well-known *Chezy's formula* for open channel flow. Chezy's formula is probably the first formula derived for uniform flow. The constant C is commonly known as Chezy's resistance factor; it was found to vary in relation to both the conditions of the channel and the flow.

Over the past two centuries many attempts have been made to determine the value of Chezy's C. The simplest relationship and the one most widely applied in the United States is derived from the work of an Irish engineer, Robert Manning (1891–1895).* Using the analysis performed on his own experimental data and on those of others, Manning derived the following empirical relation:

$$C = \frac{1}{n} R_H^{1/6} \qquad (5.3)$$

in which n is known as the Manning's coefficient of the channel roughness. Some typical values of Manning's coefficient are given in Table 5.1.

Substituting Eq. (5.3) into Eq. (5.2), we have the *Manning's formula for*

TABLE 5.1 Typical Values of Manning's *n*

Channel Surface	n
Smooth steel surface	0.012
Corrugated metal	0.024
Smooth concrete	0.011
Concrete culvert (with connection)	0.013
Glazed brick	0.013
Earth excavation, clean	0.022
Natural stream bed, clean, straight	0.030
Smooth rock cuts	0.035
Channels not maintained	0.050–0.1

*Robert Manning, "On the Flow of Water in Open Channels and Pipes," *Transactions, Institution of Civil Engineering of Ireland, 10* (1891), pp. 161–207; *24* (1895), pp. 179–207.

uniform flow:

$$V = \frac{1}{n} R_H^{2/3} S^{1/2} \tag{5.4}$$

The discharge in a uniform flow channel may be determined by

$$Q = AV = \frac{1}{n} A R_H^{2/3} S^{1/2} \tag{5.5}$$

On the right-hand side of this equation, the water area, A, and the hydraulic radius, R_H, are both functions of water depth, d, which is known as the *uniform depth* or *normal depth*.

The computation of uniform flow may be performed by the use of either Eq. (5.5) or Eq. (5.4) and involves basically six variables: the roughness coefficient, n; the channel slope, S; the normal discharge, Q; the channel geometry that includes the water area, A, and the hydraulic radius, R_H; the mean velocity, V; and the normal depth, d.

In practical applications, any four of the six variables are given, and the remaining two unknowns can be determined by using the equations.

Example 5.1

A 3-m-wide rectangular irrigation channel carries a discharge of 25.3 m³/s at a uniform depth of 1.2 m. Determine the slope of the channel with Manning coefficient, $n = 0.022$.

Solution

For a rectangular channel, $z = 0$, so that the area, the wetted perimeter and the hydraulic radius are:

$$A = (b + zd)d = bd = 3 \cdot 1.2 = 3.6 \text{ m}^2$$
$$P = b + 2\sqrt{1 + z^2}\, d = b + 2d = 5.4 \text{ m}$$
$$R_H = \frac{A}{P} = \frac{3.6}{5.4} = \frac{2}{3} = 0.667 \text{ m}$$

Equation 5.5 can be rewritten as

$$S = \left(\frac{Qn}{AR_H^{2/3}}\right)^2 = 0.041$$

Example 5.2

A trapezoidal open channel (see Figure 5.4) has a bottom width of 10 m and side slopes with inclination 1: 2. The channel is paved with smooth cement surface. If the channel is laid on a slope of 0.0001 and carries a uniform flow at the depth of 2 m, determine the discharge.

Figure 5.4

Solution

The area of the trapezoidal section is

$$A = b \cdot d + 2d \cdot \frac{zd}{2} = (b + zd)d$$

where

$$z = \frac{1}{\text{side slope}} = \frac{1}{\frac{1}{2}} = 2$$

$$A = (10 + 2 \cdot 2) \cdot 2 = 28 \text{ m}^2$$

The wetted perimeter is

$$P = b + 2(d^2 + z^2d^2)^{1/2} = b + 2\sqrt{1 + z^2}d = 18.94 \text{ m}$$

Then, the hydraulic radius is

$$R_H = \frac{A}{P} = \frac{28}{18.94} = 1.478 \text{ m}$$

Substituting the above value into Eq. (5.5), we get

$$Q = \frac{1}{n} A R_H^{2/3} S^{1/2} = \frac{1}{0.011}(28)(1.478)^{2/3}(0.0001)^{1/2}$$

$$Q = 33.03 \text{ m}^3/\text{s}$$

Example 5.3

If the discharge in the channel in Example 5.1 is increased to 40 m³/s, what is the normal depth of the flow?

Solution

The geometric parameters are

$$\text{Area: } A = bd = 3d$$

$$\text{Wetted Perimeter: } P = b + 2d = 3 + 2d$$

$$\text{Hydraulic Radius: } R_H = \frac{A}{P} = \frac{3d}{3 + 2d}$$

Substituting these values in Eq. (5.5), we get

$$Q = \frac{1}{n} A R_H^{2/3} S^{1/2}$$

$$40 = \frac{1}{0.022}(3d)\left(\frac{3d}{3+2d}\right)^{2/3}(0.041)^{1/2}$$

or

$$AR^{2/3} = (3d)\left(\frac{3d}{3+2d}\right)^{2/3} = \frac{(0.022)(40)}{(0.041)^{1/2}} = 4.346$$

Solving by trial, we find that

$$d = 1.69 \text{ m}$$

PROBLEMS

5.2.1. Determine the normal depth for the channel in Example 5.1 if the discharge were 50 m³/s.

5.2.2. A triangular channel with side slopes of 45° paved with brick in cement mortar should deliver 4 m³/s. Determine the depth of the flow if the slope is 0.0016.

5.2.3. A rectangular channel with slope 0.0004 is cut in rock. The width will be twice the depth. Determine its dimensions for a discharge of 50 m³/s.

5.2.4. Determine the discharge if the channel in Prob. 5.2.3 is paved with smooth concrete.

5.2.5. A rectangular channel 3 m wide carries a discharge of 6.3 m³/s at the normal depth of 0.8 m. The channel is laid on a slope of $S_0 = 0.01$. Determine the roughness coefficient n.

5.2.6. A rectangular chute is designed to transport lumber from a high mountain region to a lower lake. The channel must remain a uniform depth of 1 m with a discharge rate of 10 m³/s. The channel is to be built with a material with $n = 0.016$. Determine the slope of the channel if its width is 3 m.

5.2.7. A triangular highway gutter (Figure 5.5) is designed to carry a discharge of 52 m³/min on a channel slope of 0.0016. The gutter is 0.8 m deep with one side vertical and one side sloped at 1 on z. The channel surface is clean earth excavation. Determine the side slope z.

Figure 5.5

5.2.8. A trapezoidal channel with 6.2-m-wide bottom and side slope $z = 2$ carries a discharge of 10 m³/s. The channel is built with glazed brick surface on a 0.04 channel slope. Using Chezy's formula, determine the uniform flow velocity.

5.2.9. 100 m³/s are flowing in a rectangular channel 12 m wide. If the depth of the flow is 3 m, what will be the depth when the channel contracts to 2.5 m wide? Neglect losses.

5.2.10. Research in the library to find five different uniform flow formulas. List the author(s) and limitations of each formula.

5.3 HYDRAULIC EFFICIENCY OF OPEN CHANNEL SECTIONS

The uniform flow formula [Eq. (5.5)] shows that for the same cross-sectional area, A, and channel slope, S, the channel section with a larger hydraulic radius, R_H, delivers a larger discharge. It is a section of higher *hydraulic efficiency*. Since the hydraulic radius is equal to the water cross-sectional area divided by the wetted perimeter, for a given cross-section area, the channel section with the least wetted perimeter is the *best hydraulic section*.

Among all shapes of cross-sectional area, the semicircle has the least perimeter with the given area; hence, it is the most hydraulically efficient of all sections. A channel with a semicircular cross section, however, has sides that are curved and that are almost vertical at the water surface level, which makes the channel expensive in initial construction and difficult to maintain. In practice, semicircular sections are only used in smaller flumes of prefabricated materials.

For larger channels, trapezoidal sections are most commonly used. The most efficient trapezoidal section is a half-hexagon which can be inscribed to a semicircle with its center at the free water surface and 60° angle on the sides. The next commonly used channel section is the rectangular section. The most efficient rectangular section is the half-square section which can also be inscribed to a semicircle with the center of the circle at the free water surface. The hydraulically efficient semicircular, half-hexagon, and half-square sections are shown in Figure 5.6.

The concept of hydraulic efficient sections is only valid when the channel is lined with stabilized, nonerodible materials. Idealistically, a channel should

Figure 5.6. Hydraulically efficient sections.

always be designed for the best hydraulic efficiency, but it should be modified for practicability and the cost of construction. It should be noted that although the best hydraulic section offers the minimum water area for a given discharge, it does not necessarily have the minimum excavation or cost of construction. A half-hexagon section, for example, is a best hydraulic section only when the water surface reaches the level of the bank top. This section is not suitable for general applications because a sufficient distance above the water surface must be provided to prevent waves or fluctuations of water surface from overflowing the sides. The vertical distance from the designed water surface to the top of the channel banks is known as the *freeboard* of the channel.

The height of the freeboard may vary from 5% to 30% of the designed water depth, depending on the channel design and purposes. There is no accepted rule for the determination of the freeboard, since the wave actions and the fluctuations in water surface elevations may vary greatly from one channel to another. In general, the freeboard is considered related to the designed discharge of the channel. The U.S. Bureau of Reclamation provides a chart that can be used as a guide for average freeboard and bank heights for lined-channel design, as shown in Figure 5.7.

Figure 5.7. Recommended freeboard and height of banks in lined channels. (From *Linings for Irrigation Canals*, U.S. Bureau of Reclamation, 1976.)

Example 5.4

Show that the best hydraulic trapezoidal section is a half-hexagon.

Solution

The water cross-sectional area, A, and the wetted perimeter, P, of a trapezoidal section are

$$A = bd + zd^2 \tag{1}$$

and

$$P = b + 2d\sqrt{1 + z^2} \tag{2}$$

From Eq. (1), $b = A/d - zd$. This relationship is substituted into Eq. (2):

$$P = \frac{A}{d} - zd + 2d\sqrt{1 + z^2}$$

First, consider both A and z constant, and let the first derivative of P with respect to d equal zero for the minimum value of P.

$$\frac{dP}{dd} = -\frac{A}{d^2} - z + 2\sqrt{1 + z^2} = 0$$

Substituting for A from Eq. (1), we get

$$\frac{bd + zd^2}{d^2} = 2\sqrt{1 + z^2} - z$$

or

$$b = 2d(\sqrt{1 + z^2} - z) \tag{3}$$

By definition, the hydraulic radius, R_H, may be expressed as

$$R_H = \frac{A}{P} = \frac{bd + zd^2}{b + 2d\sqrt{1 + z^2}}$$

Substituting the value of b from Eq. (3) into the above equation and simplifying, we have

$$R_H = \frac{d}{2}$$

It shows that the best hydraulic trapezoidal section has a hydraulic radius equal to one-half of the water depth. Substituting Eq. (3) into Eq. (2) and solving for P, we have

$$P = 2d(2\sqrt{1 + z^2} - z) \tag{4}$$

In order to determine the value of z that makes P the least, the first derivative of P is taken with respect to z. Equating it to zero and simplifying, we have

$$z = \frac{\sqrt{3}}{3} = \cot 60°$$

and thus,

$$b = 2d\left(\sqrt{1 + \frac{1}{3}} - \frac{\sqrt{3}}{3}\right) = 2\frac{\sqrt{3}}{3}d \quad \text{or} \quad d = \frac{\sqrt{3}}{2}b = b \sin 60°$$

This means that the section is a half-hexagon (see Figure 5.8).

$$z = \frac{1}{\tan 60°} = \frac{\sqrt{3}}{3}$$

$$d = b \sin 60° = \frac{\sqrt{3}}{2}b$$

Figure 5.8

PROBLEMS

5.3.1. What is the depth-to-width ratio for the most hydraulically efficient rectangular section?

5.3.2. What would be the minimum excavated volume for a channel cut in rock (rectangular section)? It should deliver 25 m³/s in a slope of 0.003 m/m.

5.4   ENERGY PRINCIPLES IN OPEN CHANNEL FLOWS

The energy principles derived for water flow in pipes are generally applicable to open channel flows. The energy contained in a unit weight of water flowing in an open channel may also be measured in the three basic forms: kinetic energy, pressure energy, and elevation energy above a certain energy datum line.

Kinetic energy in any section of an open channel is expressed in the form of $V^2/2g$, where V is the mean velocity defined by the discharge divided by the water area, $V = Q/A$, in the section. The actual velocity of water flowing in an open channel section may vary in different parts of the section. The velocities near the channel bed are retarded due to friction and they reach a maximum near the water surface in the center part of the channel. The distribution of velocities in a cross section results in a different value of kinetic energy for each part of the section. An average value of the energy in a section of open channel may be expressed in terms of the mean velocity as $\alpha(V^2/2g)$, where α is known as the *energy coefficient*. The value of α depends on the actual velocity distribution in a particular channel section. Its value is always greater than unity. The ordinary range of α lies between 1.05 for uniformly distributed velocities and 1.20 for highly unevenly distributed velocities in a section. In simple analysis, however, the velocity heads (kinetic energy head) in an open channel are taken as $V^2/2g$ by assuming α equal to unity as an approximation.

Since open channel flow always has a free surface that is exposed to the atmosphere, the pressure on the free surface is constant and commonly taken as a zero pressure reference. Pressure energy in open channel is usually computed with reference to the free surface. If the flow in a channel is approximately along a straight-line slope, the pressure head at any submerged point A is equal to the vertical distance (the depth) measured from the free surface to the point. For a given open channel section, the water depth at the section, d, is commonly used to represent the pressure head, $p/\gamma = d$. However, if the water is flowing over a vertical curve, such as a spillway or a weir, the centrifugal force produced by the fluid mass flowing over the curved path may cause a marked difference in pressure from that directly measured from the depth. For water flows over a convex path (Figure 5.9a), the centrifugal force acts in the direction opposite to the gravity force and the pressure energy is less than that of the water depth, by mv^2/r, where m is the mass of water column immediately above a unit area and v^2/r is the centrifugal

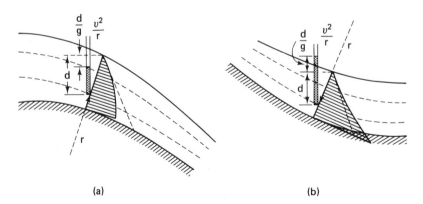

(a)   (b)

Figure 5.9. Flow over curved surfaces. (a) Convex path. (b) Concave path.

acceleration of the water mass flowing along a path with radius of curvature, r. The resulting pressure head is

$$\frac{p}{\gamma} = d - \frac{d}{g} \cdot \frac{v^2}{r} \tag{5.6}$$

For water flows over a concave path (Figure 5.9b), the centrifugal force is in the same direction as the gravity force, and the pressure energy is greater than that represented by the water depth. The resulting pressure head is

$$\frac{p}{\gamma} = d + \frac{dv^2}{gr} \tag{5.6a}$$

where γ is the unit weight of water, d is the depth measured from the free water surface to the point of interest, v is the velocity at the point, and r is the radius of curvature of the curved flow path.

The elevation energy head of open channel flow is measured with respect to a selected horizontal datum line. The vertical distance measured from the datum to the channel bottom is commoly taken as the elevation (potential) energy head at the section.

The total energy head at any section in an open channel may therefore be generally expressed as,

$$H = \frac{V^2}{2g} + d + z \tag{5.7}$$

Specific energy in a channel section is defined as the energy head measured with respect to the channel bottom at the section. According to Eq. (5.7), the specific energy at any section is

$$E = \frac{V^2}{2g} + d \tag{5.8}$$

or, the specific energy at any section in an open channel is equal to the sum of the velocity head and the water depth at the section.

Given the water area, A, and the discharge, Q, at a particular section, the above equation may be rewritten as

$$E = \frac{Q^2}{2gA^2} + d \tag{5.9}$$

Thus, for a given discharge Q, the specific energy at any section is a function of the depth of the flow only.

When the depth of the flow, d, is plotted against the specific energy for a given discharge at a given section, a *specific energy curve* is obtained (see Figure 5.10).

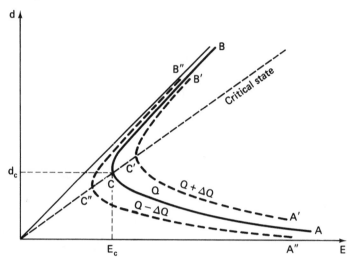

Figure 5.10. Specific energy curves for different discharges in a given open channel.

The specific energy curve has two limbs, AC and CB. The lower limb always approaches the horizontal axis toward the right and the upper limb approaches asymptotically to the 45°-line that passes through the origin.* At any point P on the specific energy curve the ordinate represents the depth of the flow at the section and the abscissa represents the corresponding specific energy. Usually, the same scales are used for both the ordinate and the abscissa.

In general, a family of similar curves may be plotted for various values of Q's at a given section. For higher discharge, the curve moves to the right, $A'C'B'$; for lower discharge, the curve moves to the left, $A''C''B''$.

The vertex C on a specific energy curve represents the depth, d_c, at which the discharge Q may be delivered through the section at minimum energy, E_c. This depth is commonly known as the *critical depth* for the discharge Q at the given section. The corresponding flow in the section is known as the *critical flow*. At a smaller depth the same discharge can be delivered only by a higher velocity and, hence, a higher specific energy. The state of rapid and shallow flow through a section is known as the *supercritical flow* or *rapid flow*. At a larger depth the same discharge may be delivered through the sec-

*For a channel of very large slope, the upper limb may approach a line with angle of inclination different from 45°.

tion with a smaller velocity but also a higher specific energy. The tranquil, high-stage flow is known as the *subcritical flow*.

At the critical state the specific energy of the flow takes a minimum value. This value can be computed by equating to zero the first derivative of the specific energy with respect to the water depth.

$$\frac{dE}{dd} = \frac{d}{dd}\left(\frac{Q^2}{2gA^2} + d\right) = -\frac{Q^2}{gA^3}\frac{dA}{dd} + 1 = 0$$

The differential water area dA/dd near the free surface is $dA/dd = T$, where T is the top width of the channel section. Hence,

$$-\frac{Q^2T}{gA^3} + 1 = 0$$

An important parameter for open channel flow is defined by $A/T = D$, which is known as the *hydraulic depth* of the section. For rectangular cross sections, the hydraulic depth is equal to the depth of the flow. The above equation may thus be simplified to

$$\frac{dE}{dd} = 1 - \frac{Q^2}{gDA^2} = 1 - \frac{V^2}{gD} \tag{5.10}$$

or

$$\frac{V}{\sqrt{gD}} = 1 \tag{5.11}$$

The quantity $V/\sqrt{gD}$ is dimensionless. It can be derived as the ratio of the inertial force in the flow to the gravity force in the flow (see Chapter 8 for detailed discussion). This ratio may also be interpreted physically as the ratio between the mean flow velocity, V, and the speed of a small gravity (surface) wave traveling over the water surface. It is known as the *Froude number*, N_F.

$$N_F = \frac{V}{\sqrt{gD}} \tag{5.12}$$

When the Froude number is equal to unity, as indicated by Eq. (5.11), $V = \sqrt{gD}$, the speed of the surface wave (disturbance wave) and that of the flow is the same. The flow is in the critical state. When the Froude number is less than unity, $V < \sqrt{gD}$, the flow velocity is smaller than the speed of a surface wave traveling on the water surface. The flow is in the subcritical state. When the Froude number is greater than unity, $V > \sqrt{gD}$, the flow is in the supercritical state.

From Eq. 5.10 we may also write for critical flow,

$$\frac{Q^2}{g} = DA^2$$

In a rectangular channel, $D = d$ and $A = bd$, we have

$$\frac{Q^2}{g} = d^3 b^2$$

Since this relation is derived from the critical flow conditions stated above, $d = d_c$, which is the critical depth, and

$$d_c = \sqrt[3]{\frac{Q^2}{gb^2}} = \sqrt[3]{\frac{q^2}{g}} \tag{5.13}$$

where q is the discharge per unit width of the channel. For an open channel of any sectional shape, the critical depth is always a function of the channel discharge and does not vary with the change of the channel slope.

Example 5.5

A hydraulic transition is designed to connect two rectangular channels of the same width by a sloped floor (Figure 5.11). If the channel is 3 m wide and is carrying a discharge of 15 m³/s at 3.6 m depth, determine the water surface profile in the transition. Assume 0.1 m energy loss uniformly distributed through the transition.

Solution

The specific energy curve can be constructed based on the given discharge and the sectional geometry by using the following relationship from Eq. (5.9).

$$E = \frac{Q^2}{2gA^2} + d = \frac{(15)^2}{2 \cdot 9.81 \cdot (3d)^2} + d = \frac{1.27}{d^2} + d$$

At the inlet to the transition the velocity is V_i:

$$V_i = \frac{Q}{A_i} = \frac{15}{3.6 \cdot 3} = 1.39 \text{ m/s}$$

Where A_i is the water area at the inlet and the velocity head is

$$\frac{V_i^2}{2g} = \frac{1.39^2}{2 \cdot 1.98} = 0.10 \text{ m}$$

The total energy head at the inlet as measured with respect to the datum

(a)

$$d_c = \sqrt{\frac{5^2}{g}} = 1.37 \text{ m}$$

$$t_c = 2.05 \text{ m}$$

d (m)	E (m)
0.5	5.60
1.0	2.27
2.0	2.32
3.0	3.14
4.0	4.07

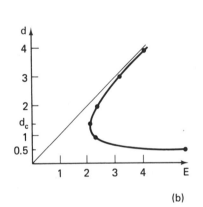

(b)

Figure 5.11

line is

$$H_i = \frac{V_i^2}{2g} + d_i + z_i = 0.10 + 3.60 + 0.4 = 4.10 \text{ m}$$

The top horizontal line in Figure 5.11 shows this energy level.

At the exit of the transition the total energy available is less by 0.1 m, as indicated by the energy gradient line (EGL) in Figure 5.11a.

$$H_e = \frac{V_e^2}{2g} + d_e + z_e = H_i - 0.1 = 4.00 \text{ m}$$

E_e is the specific energy measured with respect to the channel bottom

$$E_e = H_e = 4.00 \text{ m}$$

This value is applied to the specific energy curve (Figure 5.11) to obtain the water depth at the exit section, as shown.

Water surface elevations at four other sections 4.00 m; 8.00 m; 12.00 m; 16.00 m from the entrance section are computed by using the same method. The results for all six sections are tabulated below:

Section	Inlet	4.00 m	8.00 m	12.00 m	16.00 m	Exit
Specific Energy E (m)	3.70	3.76	3.82	3.88	3.94	4.00
Water Depth d (m)	3.60	3.665	3.729	3.792	3.855	3.917

PROBLEMS

5.4.1. A trapezoidal channel with a bottom width of 4 m and side slopes of $z = 1.5$ is carrying a discharge of 50 m³/s at a depth of 3 m. Determine the following:

(a) The alternate depth for the same specific energy.

(b) The critical depth.

(c) The uniform flow depth for a slope of 0.0004 and $n = 0.022$.

5.4.2. A 15-m-wide rectangular channel with channel slope $S = 0.0025$ and Manning's coefficient $n = 0.035$ is carrying a discharge of 62 m³/s. Determine the normal depth and the critical depth of the flow. Construct the specific energy curve.

5.4.3. A 3-m-wide rectangular channel carries a discharge 15 m³/s at a uniform depth of 1.7 m. The Manning's coefficient $n = 0.022$. Determine (a) the channel slope, (b) the critical depth, and (c) the Froude number.

5.4.4. At the uniform depth of 1.2 m a trapezoidal channel has a bottom width of 1.5 m and side slopes $z = 1.0$. The channel has $n = 0.025$ and slope $S = 0.004$. Determine the discharge and calculate the speed of the surface wave.

5.4.5. A flow of 100 m³/s occurs in a trapezoidal canal having a bottom width of 10 m, a side slope $z = 2.0$, and $n = 0.017$. Calculate the critical depth and critical slope for the flow.

5.4.6. A 12.5-m side rectangular channel carries 32 m³/s at the depth of 2 m. Is this flow subcritical or supercritical? If $n = 0.025$, what is the critical slope of this channel for this discharge? What channel slope must be provided to produce a uniform flow at the depth of 2 m?

5.4.7. Repeat Example 5.5 for an inlet depth of 1.50 m.

5.4.8. A transition is constructed to connect two trapezoidal channels with bottom slopes of 0.001 and 0.0004, respectively. The channels have the same cross-sectional shape, with a bottom width of 3 m, a side slope of $z = 2$, and Manning's coefficient $n = 0.02$. The transition is 20 m long and is designed to carry a discharge of 20 m³/s. Assume that the transition has the same kind of surface as that of the channels and that an energy loss of 0.02 m is uniformly distributed through the transitional length. Determine the change in the bottom elevations at the two ends of the transition.

5.4.9. A hydraulic transition 30 m long is used to connect two rectangular channels, 4 m and 2 m wide, respectively. The design discharge is 18 m³/s, $n = 0.013$, and the slopes of the channels are 0.0009 for both. Determine the change in the bottom elevation and the water surface profile in the transition if the energy loss in the transition is 0.5 m and is uniformly distributed throughout the transitional length. Assume uniform flows before and after the transition.

5.5 HYDRAULIC JUMPS

Hydraulic jump is a natural phenomenon in open channel. It is an abrupt reduction in flow velocity by means of a sudden increase of water depth in the downstream direction.

Through a hydraulic jump, a high-velocity supercritical flow (upstream) is changed to a low-velocity subcritical flow (downstream). Correspondingly, a low-stage supercritical depth, d_1, is changed to a high-stage subcritical depth, d_2, known, respectively, as the *initial depth* and the *sequent depth* of a hydraulic jump. In the region of the hydraulic jump the characteristic rolling of surface and boiling of water can be seen. These violent motions are accompanied by a significant loss of energy head through the jump. Given the discharge in a particular channel, the amount of loss of energy head through a jump, ΔE, can be determined by simply measuring the initial and sequent depths and using the specific energy curve as shown in Figure 5.12. Predicting the sequential depth by estimating the energy loss, however, is impractical because it is difficult to determine the energy loss through a jump. The relationship between the initial depth and the sequent depth in a hydraulic jump may be determined by considering the balance of forces and momentum immediately before and after the jump.

Take the free body of the water volume involved in the vicinity of a jump, as shown in Figure 5.12. The balance between the hydrostatic force F_1 and

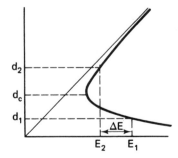

Figure 5.12. Hydraulic jump.

F_2, represented by the two triangles and the momentum flux through section 1 and section 2, per unit width of the channel may be expressed as

$$F_1 - F_2 = \rho q(V_2 - V_1) \tag{5.14}$$

Substituting the following quantities

$$F_1 = \frac{\gamma}{2}d_1^2, \qquad F_2 = \frac{\gamma}{2}d_2^2, \qquad V_1 = \frac{q}{d_1}, \qquad V_2 = \frac{q}{d_2}$$

into Eq. (5.14) and simplifying, we get

$$\frac{q^2}{g} = d_1 d_2 \left(\frac{d_1 + d_2}{2}\right) \tag{5.15}$$

where q is the discharge per unit width of the channel. This equation may also be rearranged into a more convenient form as follows:

$$\frac{d_2}{d_1} = \frac{1}{2}(\sqrt{1 + 8N_{F_1}^2} - 1) \tag{5.16}$$

where N_{F_1} is the Froude number of the approaching flow,

$$N_{F_1} = \frac{V_1}{\sqrt{gd_1}} \tag{5.17}$$

Example 5.6

A 3-m-wide rectangular channel carries 15 m³/s of water at 0.6 m depth before entering a jump. Compute the downstream water depth and the critical depth.

Solution

The discharge per unit width is:

$$q = \frac{15}{3} = 5 \text{ m}^3/\text{s/m}$$

The critical depth, by Eq. (5.13), is:

$$d_c = \sqrt[3]{\frac{5^2}{9.81}} = 1.60 \text{ m}$$

The approaching velocity is:

$$V_1 = \frac{q}{d_1} = \frac{5}{0.6} = 8.33 \text{ m/s}$$

The Froude number for the approaching flow can be computed by using this velocity and the initial depth $d_1 = 0.6$.

$$N_{F_1} = \frac{V_1}{\sqrt{gd_1}} = 3.43$$

Substituting this value into Eq. (5.16) gives

$$\frac{d_2}{0.6} = \frac{1}{2}(\sqrt{1 + 8(3.43)^2} - 1)$$

Solve for the sequent depth d_2.

$$d_2 = 4.38 \text{ m}$$

Equation (5.14) may also be arranged to

$$F_1 + \rho q V_1 = F_2 + \rho q V_2$$

or

$$F_s = F + \rho q V \tag{5.18}$$

The quantity F_s is known as the *specific force* per unit width of the channel. For a given discharge, the specific force is a function of the water depth at a given section. When the specific force, F_s, is plotted against the water depth, it also has two limbs and a vertex that appears at the critical depth. A typical specific force curve is shown in Figure 5.13.

Hydraulic jump usually takes place in a rather short reach in a channel. Therefore, it is reasonable to assume that through a hydraulic jump the

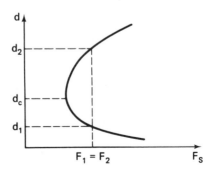

Figure 5.13. Specific force curve.

specific forces immediately before and after a jump are approximately the same. The value of F_s can be computed from the given conditions of the approaching flow. If we apply this value to the specific force curve in Figure 5.13, we can draw a vertical line that gives both the initial and sequent depths of a jump.

The energy head loss through the hydraulic jump, ΔE, may then be estimated by applying the definition

$$\Delta E = \left(\frac{V_1^2}{2g} + d_1\right) - \left(\frac{V_2^2}{2g} + d_2\right)$$

$$= \frac{1}{2g}(V_1^2 - V_2^2) + (d_1 - d_2) = \frac{q^2}{2g}\left(\frac{1}{d_1^2} - \frac{1}{d_2^2}\right) + (d_1 - d_2)$$

Substituting Eq. (5.15) into the above equation and simplifying, we get

$$\Delta E = \frac{(d_2 - d_1)^3}{4d_1 d_2} \tag{5.19}$$

Example 5.7

A long rectangular open channel 3 m wide carries a discharge of 15 m³/s. The channel slope is 0.004 and the Manning's coefficient is 0.01. At a certain point in the channel where the flow reaches the normal depth,

(a) Determine the state of the flow. Is it supercritical or subcritical?
(b) If a hydraulic jump takes place at this depth, what is the sequent depth at the jump?
(c) Estimate the energy head loss through the jump.

Solution

(a) The critical depth for the flow was calculated in Example 5.6, where $d_c = 1.60$ m. The normal depth of this channel can be determined by the Manning equation, Eq. (5.5):

$$Q = \frac{1}{n} A_1 R_{H_1}^{2/3} S^{1/2}$$

where

$$A = d_1 b, \qquad R_H = \frac{A_1}{P_1} = \frac{d_1 b}{2d + b}, \qquad b = 3 \text{ m}$$

We have

$$15 = \frac{1}{0.01}(3d_1)\left(\frac{3d_1}{2d_1 + 3}\right)^{2/3}(0.004)^{1/2}$$

Solving the equation for d_1, we get

$$d_1 = 1.08 \text{ m}, \qquad V_1 = \frac{15}{3d_1} = 4.63 \text{ m/s}$$

and

$$N_{F_1} = \frac{V_1}{\sqrt{gd_1}} = 1.42$$

As $N_{F_1} > 1$, the flow is supercritical.

(b) Applying Eq. 5.16, we get

$$d_2 = \frac{d_1}{2}(\sqrt{1 + 8N_{F_1}^2} - 1) = 1.57d_1 = 1.70 \text{ m}$$

(c) The head loss can be estimated by using Eq. (5.19).

$$\Delta E = \frac{(d_2 - d_1)^3}{4d_2d_1} = \frac{(0.62)^3}{4 \cdot 1.7 \cdot 1.08} = 0.032 \text{ m}$$

PROBLEMS

5.5.1. A hydraulic jump occurs at an initial depth of 1.5 m in a rectangular channel 4 m wide. If the sequent depth is 2.2 m, determine the energy loss and the discharge in the channel.

5.5.2. Is the flow in the channel of Prob. 5.4.3 subcritical or supercritical? Determine the sequent depth and the energy loss if a hydraulic jump takes place at a depth of 1 m.

5.5.3. Discuss how the change in discharge would affect the specific force curve.

5.5.4. Construct the specific energy and specific force curves for a circular pipe of 1-m diameter that carries 1 m^3/s.

5.5.5. Discuss why the specific energy curve usually approaches the 45°-line as a limit.

5.5.6. Plot the specific force diagram for a 10-m rectangular channel carrying 15 m^3/s discharge. Determine the critical depth and the minimum specific energy.

5.6   GRADUALLY VARIED FLOWS

Gradually varied flows in open channels differ from the uniform flows and the rapidly varied flows (hydraulic jumps, flow through a streamlined transition, etc.), in that the changes in water depth in the channel take place very gradually with distance.

In uniform flow the water depth remains a constant value known as the normal depth (or uniform depth). The energy gradient line is parallel to the water surface line and the channel bottom line. The velocity distribution remains unchanged throughout the reach. Thus, the computation of only one water depth is necessary for the entire reach.

In rapidly varied flows, such as hydraulic jumps, rapid changes in water depth take place in a short distance. A significant change in water velocities is associated with the rapid variation of water cross-sectional area. At this high rate of flow deceleration, the energy loss is inevitably high. The computation of water depths using the energy principles is not reliable. In this case, computations can only be carried out by applying the momentum principles [Eq. (5.14)].

In gradually varied flows the velocity changes take place very gradually with distance, so that the effects of acceleration on the flow between two adjacent sections are negligible. Computation of the water surface profiles can be carried out based strictly on energy considerations.

The total energy head at any section in an open channel can be expressed as in Eq. (5.7),

$$H = \frac{V^2}{2g} + d + z = \frac{Q^2}{2gA^2} + d + z$$

In order to compute the water surface profile, it is necessary to obtain first the variation of the total energy head along the channel. Differentiating H with respect to the channel distance x, we obtain the energy gradient in the direction of the flow.

$$\frac{dH}{dx} = \frac{-Q^2}{gA^3}\frac{dA}{dx} + \frac{dd}{dx} + \frac{dz}{dx} = -\frac{Q^2 T}{gA^3}\frac{dd}{dx} + \frac{dd}{dx} + \frac{dz}{dx}$$

where $dA = T(dd)$. Rearranging the equation gives

$$\frac{dd}{dx} = \frac{\dfrac{dH}{dx} - \dfrac{dz}{dx}}{1 - \dfrac{Q^2 T}{gA^3}} \tag{5.20}$$

The term dH/dx is the slope of the energy gradient line. It is always a negative quantity since the total energy head reduces in the direction of the flow, or $S_e = -dH/dx$. Similarly, the term dz/dx is the slope of the channel bed. It is negative when the elevation of the channel bed reduces in the direction of the flow; it is positive when the elevation of the channel bed increases in the direction of the flow. In general, we may write $S_0 = -dz/dx$.

The energy slope in gradually varied flow between two adjacent stations may also be approximated by using a uniform flow formula. For simplicity, the derivation will be demonstrated with a wide rectangular channel section where $A = bd$, $Q = bq$, and $R_H = A/P = bD/(b + 2D) \cong D = d$ $(b \gg d$ in wide rectangular channels).

Using the Manning formula [Equation (5.5)], we get

$$S_e = -\frac{dH}{dx} = \frac{n^2 Q^2}{R_H^{4/3} A^2} = \frac{n^2 Q^2}{b^2 d^{10/3}} \tag{5.21}$$

The slope of the channel bed may also be expressed in similar terms if uniform flow were assumed to take place in the channel. Since in uniform flows the slope of the channel bed is equal to the energy slope, the hypothetical uniform flow conditions are designated with the subscript n. We have

$$S_0 = \frac{-dz}{dx} = \left(\frac{n^2 Q^2}{b^2 d^{10/3}}\right)_n \tag{5.22}$$

From Eq. (5.13)

$$d_c = \sqrt[3]{\frac{q^2}{g}} = \sqrt[3]{\frac{Q^2}{gb^2}} .$$

or

$$Q^2 = gd_c^3 b^2 = \frac{gA_c^3}{b} \tag{5.23}$$

Substituting now Eqs. (5.21), (5.22), and (5.23) into Eq. (5.20), we have

$$\frac{dd}{dx} = \frac{S_0 \left[1 - \left(\frac{d_n}{d}\right)^{10/3}\right]}{\left[1 - \left(\frac{d_c}{d}\right)^3\right]} \tag{5.24}$$

This is the general differential equation for gradually varied flows. It is also referred to as the *gradually varied flow equation*. The term dd/dx represents the slope of the water surface with respect to the bottom of the channel. For $dd/dx = 0$, the water depth remains a constant value throughout the reach. It represents the special case of a uniform flow. For $dd/dx < 0$, the water depth decreases in the direction of the flow. For $dd/dx > 0$, the water depth increases in the direction of the flow. The solutions of this equation under given conditions will provide the various water surface profiles that may occur in open channels.

5.7 CLASSIFICATION OF GRADUALLY VARIED FLOWS

In analyzing the gradually varied flows the role of critical depth, d_c, is very important. When open channel flow approaches the critical depth ($d = d_c$), the denominator of Eq. (5.24) approaches zero and the value of dd/dx approaches infinity. The water surface becomes very steep. This is seen at hydraulic jumps or at a water surface entering a steep channel from a mild channel or a lake. The latter provides a unique one-to-one relationship between the discharge and the water depth in a channel and is known as the control of open channel flows.

Depending on the channel slope, the surface conditions, the sectional geometry, and the discharge, open channels may be classified into five categories: steep channels, critical channels, mild channels, horizontal channels, and adverse channels. They are categorized according to the flow conditions in the channel as indicated by the relative positions of the normal depth, d_n, and the critical depth, d_c, calculated for each particular channel. The criteria are as follows:

$$\text{Steep channels: } d_n/d_c < 1.0 \quad \text{or} \quad d_n < d_c$$
$$\text{Critical channels: } d_n/d_c = 1.0 \quad \text{or} \quad d_n = d_c$$
$$\text{Mild channels: } d_n/d_c > 1.0 \quad \text{or} \quad d_n > d_c$$
$$\text{Horizontal channels: } \quad S_0 = 0$$
$$\text{Adverse channels: } \quad S_0 < 0$$

A further classification of water surface profile curves depends on the actual water depth and its relations to the critical and normal depths. The ratios of d/d_c and d/d_n may be used in the analysis, where d is the actual water depth at any section of interest in the channel.

If both d/d_c and d/d_n are greater than 1.0, the water surface profile curve is above both the critical depth line and the normal depth line in the channel (see Figure 5.14). The curve is designated as a type-1 curve. There are S-1, C-1, and M-1 curves for steep, critical, and mild channels, respectively.

If the water depth d is between the normal depth and the critical depth, the curves are designated as type-2 curves. There are S-2, M-2, H-2, and A-2 curves. The type-2 curve does not exist in critical channels. In critical channels the normal depth is the critical depth, and no distance is allowed between the normal depth line and the critical depth line.

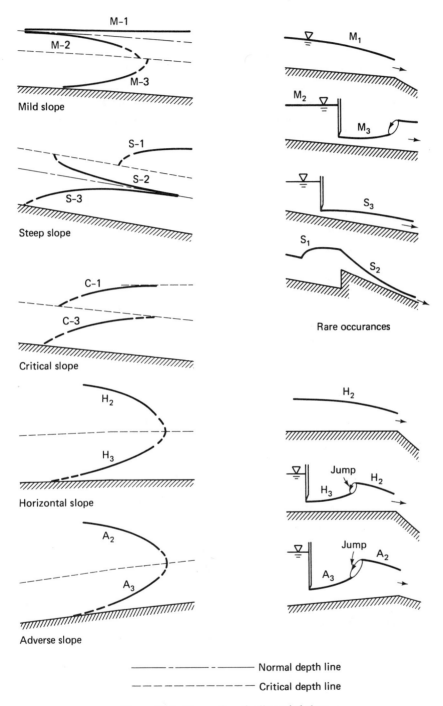

Figure 5.14. Types of gradually varied slope.

If the water depth d is less than both d_c and d_n, the water surface profile curves are type-3. There are S-3, C-3, M-3, H-3, and A-3 curves. Each of these water surface profile curves is listed and shown schematically in Figure 5.14. Examples of physical occurrence in open channels are also given.

Certain important characteristics of the water surface profile curves can be demonstrated from direct analysis of the gradually varied flow equation [Eq. (5.24)].

1. For type-1 curves, $d/d_c > 1$ and $d/d_n > 1$, the value of dd/dx is positive, indicating that water depth increases in the direction of the flow.

2. For type-2 curves, the value of dd/dx is negative. The water depth decreases in the direction of the flow.

3. For type-3 curves, the value of dd/dx is again positive. The water depth increases in the direction of the flow.

4. When the water depth approaches the critical depth, $d = d_c$, Eq. (5.24) gives $dd/dx = \infty$, indicating that the slope of the water surface profile curve is theoretically vertical. Hydraulic jump is a typical example.

5. A few types of water surface profile curves never approach the horizontal line (S-2, S-3, M-2, M-3, C-3, and A-3). Others approach the horizontal line asymptotically, except for the C-1 curve, which is horizontal throughout the channel reach. Since in a critical channel $d_n = d_c$, Eq. (5.24) gives $dd/dx = S_0$, indicating that the water depth increases at the same rate as the channel bed elevation decreases, which, theoretically, results in a horizontal water surface profile.

6. In steep channels, where $d_n < d_c$, the velocity of water flow is normally greater than that of the disturbance wave. For this reason, the flow conditions in the downstream channel will not affect those upstream. The change of water depth due to any channel disturbance propagates only in the downstream direction. Computation of the water surface profile should be carried out in the downstream direction.

7. In mild channels, where $d_n > d_c$, the speed of wave propagation is greater than the velocity of water flow. Any disturbance in the downstream channel can travel upstream and affects the flow conditions upstream as well as downstream. Any change of water depth in the downstream channel propagates upstream and may also change the water depth in the upstream channel. Computation of the water surface profile should be carried out in the upstream direction.

8. At the break of the channel from mild to steep slope, or a significant drop of the channel bottom, the critical depth is assumed to take place in the immediate vicinity of the brink. At this point, a definite depth-

discharge relationship can be obtained, and is frequently used as a
control section for water surface profile computations.

Table 5.2 provides a summary of the water surface profile curves.

TABLE 5.2   Characteristics of Water Surface Profile Curves

Channel	Symbol	Type	Slope	Depth	Curve
Mild	M	1	$S_0 > 0$	$d > d_n > d_c$	M-1
Mild	M	2	$S_0 > 0$	$d_n > d > d_c$	M-2
Mild	M	3	$S_0 > 0$	$d_n > d_c > d$	M-3
Critical	C	1	$S_0 > 0$	$d > d_n = d_c$	C-1
Critical	C	3	$S_0 > 0$	$d_n = d_c > d$	C-3
Steep	S	1	$S_0 > 0$	$d > d_c > d_n$	S-1
Steep	S	2	$S_0 > 0$	$d_c > d > d_n$	S-2
Steep	S	3	$S_0 > 0$	$d_c > d_n > d$	S-3
Horizontal	H	2	$S_0 = 0$	$d > d_c$	H-2
Horizontal	H	3	$S_0 = 0$	$d_c > d$	H-3
Adverse	A	2	$S_0 < 0$	$d > d_c$	A-2
Adverse	A	3	$S_0 < 0$	$d_c > d$	A-3

5.8   COMPUTATION OF WATER SURFACE PROFILES

Water surface profiles in gradually varied flows may be computed by using
Eq. (5.24). The computation normally begins from a section where the relationship between the water surface elevation (depth) and the discharge is
known. These sections are commonly known as the *control sections*. A few
examples of the control sections in open channels are demonstrated in
Figure 5.15.

A successive computational procedure is used to compute the water surface elevation at the next section at a distance short enough so that the water
surface between the two sections may be approximated by a straight line. A
step procedure is carried out in the downstream direction for rapid (supercritical) flows and in the upstream direction for tranquil (subcritical) flows.

Two commonly used computation methods are discussed in this section:
(1) the *direct integration method* and (2) the *direct step method*. These two
methods are selectively presented here to demonstrate the necessary considerations and procedures involved in computing a water surface profile.
Other methods are frequently used in hydraulic engineering practice, but
they are not discussed here because thier programming and computations are
so complex. For these other methods of computation, the readers may refer
to the textbook by V. T. Chow.*

*V. T. Chow, *Open Channel Hydraulics* (New York: McGraw-Hill, 1959).

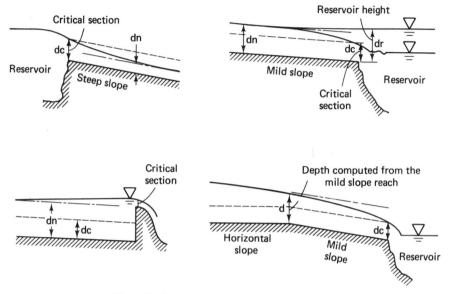

Figure 5.15. Control sections in open channels.

Direct Integration Method For over a century, engineers have made attempts to obtain a direct integration of the gradually varied flow equation [e.g., Eq. (5.24)]. Because there is such a great variety in channel cross-sectional shapes, a general solution is impossible in practice. A comprehensive review of many of the existing integration methods was made by V. T. Chow in 1955.* The method described here is suggested by Chow as an outcome of his study of the various existing methods. Chow's method can be applied directly to most of the channel shapes. In this chapter we shall discuss its application to wide rectangular channels.

The general form of the gradually varied flow equation may be rearranged and written as

$$\frac{dd}{dx} = S_0 \left[\frac{1 - \left(\frac{d_n}{d}\right)^N}{1 - \left(\frac{d_c}{d}\right)^M} \right] \tag{5.25}$$

where the exponents N and M depend on the shape of the cross section and on the flow conditions. For wide rectangular channels, $N = 10/3$ and $M = 3$, as shown in Eq. (5.24). For gradually varied flow in channels of relatively simple geometries, the values of N and M may also be assumed to be constant within the range of the integration limits.

*Chow, V. T., "Investigating the Equation of Gradually Varied Flow," Paper 838, *ASCE Proc.*, *81* (1955), 1–32. Discussions by C. J. Keifer et al. Closing discussion by V. T. Chow, Paper 1177, *ASCE Proc.*, *83*, HY1, (1957), 9–22.

By letting $d/d_n = u$, the above equation may be written as

$$dx = \frac{d_n}{S_0}\left[1 - \frac{1}{1 - u^N} + \left(\frac{d_c}{d_n}\right)^M\left(\frac{u^{N-M}}{1 - u^N}\right)\right]du$$

Hence,

$$x = \frac{d_n}{S_0}\left[u - \int_0^u \frac{du}{1 - u^N} + \left(\frac{d_c}{d_n}\right)^M \int_0^u \frac{u^{N-M}}{1 - u^N}\,du\right] + \text{constant} \quad (5.26)$$

The first integral term of the right-hand side may be represented as a function of u and N, or

$$F(u, N) = \int_0^u \frac{du}{1 - u^N} \quad (5.27)$$

which is known as the *varied flow function*.

The second term may also be expressed in similar form by letting $v = u^{N/J}$ and

$$J = \frac{N}{N - M + 1}$$

Substituting the new variables and transforming the integral into

$$\int_0^u \frac{u^{N-M}}{1 - u^N}\,du = \frac{J}{N}\int_0^v \frac{dv}{1 - v^J} = \frac{J}{N}F(v, J) \quad (5.28)$$

where

$$F(v, J) = \int_0^v \frac{dv}{1 - v^J} \quad (5.29)$$

The similarity between Eqs. (5.27) and (5.29) is obvious, except that the variables u and N are replaced by v and J in Eq. (5.29).

Substituting Eqs. (5.27) and (5.29) into Eq. (5.26), we may write

$$x = \frac{d_n}{S_0}\left[u - F(u, N) + \left(\frac{d_c}{d_n}\right)^M\left(\frac{J}{N}\right)F(v, J)\right] + \text{constant} \quad (5.30)$$

For wide rectangular channels, $N = 10/3$, $M = 3$, $J = N/(N - M + 1) = 5/2$, and $v = u^{N/J} = u^{4/3}$. Equation (5.30) may be simplified to

$$x = A\left[u - F\left(u, \frac{10}{3}\right) + BF\left(v, \frac{5}{2}\right)\right] + \text{constant} \quad (5.31)$$

where

$$A = \frac{d_n}{S_0}, \qquad B = \left(\frac{d_c}{d_n}\right)^M \left(\frac{J}{N}\right) = \left(\frac{d_c}{d_n}\right)^3 \left(\frac{3}{4}\right)$$

are constant when the discharge and the channel slope are given. $F(u, 10/3)$ and $F(v, 5/2)$, or simply $F(u, 3.33)$ and $F(v, 2.5)$, are the varied-flow functions for wide rectangular channels.

By Eq. (5.31), the length of flow profile between two consecutive sections 1 and 2 is equal to

$$\begin{aligned} L = x_2 - x_1 \\ = A\{(u_2 - u_1) - [F(u_2, 3.33) - F(u_1, 3.33)] \qquad (5.32) \\ + B[F(v_2, 2.5) - F(v_1, 2.5)]\} \end{aligned}$$

The numerical values for the varied-flow functions $F(u, 3.33)$ and $F(v, 2.50)$ for both positive channel slopes ($S_0 > 0$) and negative channel slopes ($S_0 < 0$) are provided in Table 5.3.

Example 5.8

A wide rectangular channel carries a discharge per unit width of 2.5 m³/s/m on a 0.001 slope and $n = 0.025$. Compute the backwater curve created by a low dam that has water at a depth of 2 m immediately behind the dam. The upstream computation may be carried out to a depth 1% greater than the normal depth.

Solution

The normal and critical depth for this channel can be calculated by using Eqs. (5.5) and (5.11), respectively. As for a wide rectangular channel $R_H \approx d$, and $D = d$, we may write, rearranging the Manning formula, Eq. (5.5),

$$AR_H^{2/3} = bd_n\,(d_n)^{2/3} = \frac{Q \cdot n}{S^{1/2}}$$

or

$$d_n^{5/3} = \frac{Q}{b}\frac{n}{S^{1/2}} = \frac{qn}{S^{1/2}} = \frac{2.5 \cdot 0.025}{(0.001)^{1/2}} = 1.98$$

The normal depth is

$$d_n = (1.98)^{3/5} = 1.50 \text{ m}$$

and applying the continuity condition,

$$V_c^2 = \frac{Q^2}{A_c^2} = \frac{Q^2}{b^2 d_c^2} = \frac{q^2}{d_c^2} \tag{a}$$

TABLE 5.3 The Varied-flow Functions for Wide Rectangular Channels*

$$F(x, y) = \int_0^x \frac{dx}{1 - x^y}$$

	Positive Slope			Negative Slope	
y \ x	2.50	3.33	y \ x	2.50	3.33
0.00	0.000	0.000	0.00	0.000	0.000
0.02	0.020	0.020	0.02	0.020	0.020
0.04	0.040	0.040	0.04	0.040	0.040
0.06	0.060	0.060	0.06	0.060	0.060
0.08	0.080	0.080	0.08	0.080	0.080
0.10	0.100	0.100	0.10	0.100	0.100
0.12	0.120	0.120	0.12	0.120	0.120
0.14	0.140	0.140	0.14	0.140	0.140
0.16	0.160	0.160	0.16	0.160	0.160
0.18	0.181	0.180	0.18	0.180	0.180
0.20	0.201	0.200	0.20	0.199	0.200
0.22	0.221	0.220	0.22	0.219	0.220
0.24	0.242	0.240	0.24	0.238	0.240
0.26	0.262	0.261	0.26	0.257	0.260
0.28	0.283	0.281	0.28	0.276	0.279
0.30	0.304	0.301	0.30	0.295	0.298
0.32	0.325	0.322	0.32	0.314	0.318
0.34	0.347	0.342	0.34	0.333	0.338
0.36	0.368	0.363	0.36	0.352	0.357
0.38	0.390	0.383	0.38	0.370	0.376
0.40	0.412	0.404	0.40	0.388	0.395
0.42	0.435	0.425	0.42	0.406	0.413
0.44	0.458	0.447	0.44	0.425	0.433
0.46	0.480	0.469	0.46	0.442	0.452
0.48	0.504	0.490	0.48	0.460	0.470
0.50	0.528	0.513	0.50	0.477	0.488
0.52	0.553	0.535	0.52	0.492	0.506
0.54	0.578	0.558	0.54	0.510	0.524
0.56	0.604	0.580	0.56	0.525	0.542
0.58	0.630	0.605	0.58	0.542	0.560
0.60	0.658	0.629	0.60	0.558	0.577
0.61	0.672	0.642	0.61	0.566	0.585
0.62	0.686	0.656	0.62	0.575	0.593
0.63	0.699	0.666	0.63	0.582	0.602
0.64	0.715	0.680	0.64	0.589	0.619
0.65	0.729	0.692	0.65	0.595	0.617
0.66	0.746	0.705	0.66	0.603	0.624
0.67	0.761	0.719	0.67	0.610	0.633
0.68	0.777	0.733	0.68	0.618	0.641
0.69	0.743	0.746	0.69	0.625	0.659
0.70	0.800	0.760	0.70	0.633	0.657
0.71	0.828	0.776	0.71	0.640	0.664
0.72	0.845	0.791	0.72	0.647	0.672
0.73	0.863	0.806	0.73	0.654	0.680
0.74	0.882	0.822	0.74	0.661	0.687
0.75	0.901	0.838	0.75	0.668	0.694
0.76	0.921	0.855	0.76	0.675	0.702
0.77	0.941	0.872	0.77	0.681	0.708
0.78	0.963	0.889	0.78	0.687	0.715
0.79	0.984	0.907	0.79	0.693	0.722

TABLE 5.3 The Varied-flow Functions for Wide Rectangular Channels (Continued)

$$F(x, y) = \int_0^x \frac{dx}{1 - x^y}$$

	Positive Slope			Negative Slope	
y x	2.50	3.33	y x	2.50	3.33
0.80	1.008	0.926	0.80	0.699	0.732
0.81	1.032	0.946	0.81	0.705	0.736
0.82	1.056	0.966	0.82	0.712	0.743
0.83	1.083	0.987	0.83	0.717	0.749
0.84	1.110	1.010	0.84	0.724	0.755
0.85	1.140	1.033	0.85	0.730	0.767
0.86	1.170	1.058	0.86	0.736	0.768
0.87	1.203	1.084	0.87	0.742	0.774
0.88	1.238	1.112	0.88	0.748	0.780
0.89	1.277	1.143	0.89	0.753	0.786
0.90	1.318	1.176	0.90	0.759	0.797
0.91	1.364	1.211	0.91	0.765	0.798
0.92	1.414	1.251	0.92	0.771	0.804
0.93	1.470	1.295	0.93	0.777	0.810
0.94	1.535	1.345	0.94	0.782	0.816
0.950	1.612	1.404	0.950	0.788	0.821
0.960	1.700	1.475	0.960	0.793	0.826
0.970	1.822	1.566	0.970	0.798	0.830
0.975	1.897	1.623	0.975	0.800	0.832
0.980	1.988	1.693	0.980	0.803	0.834
0.985	2.110	1.782	0.985	0.806	0.837
0.990	2.272	1.906	0.990	0.809	0.840
0.995	2.549	2.119	0.995	0.811	0.844
0.999	3.194	2.608	1.000	0.813	0.846
1.000	∞	∞	1.005	0.815	0.849
1.001	2.767	1.932	1.010	0.817	0.851
1.005	2.146	1.445	1.015	0.819	0.854
1.010	1.870	1.236	1.020	0.823	0.857
1.015	1.709	1.115	1.03	0.827	0.861
1.020	1.595	1.030	1.04	0.833	0.866
1.03	1.436	0.910	1.05	0.836	0.871
1.04	1.322	0.826	1.06	0.841	0.875
1.05	1.273	0.762	1.07	0.846	0.880
1.06	1.167	0.710	1.08	0.852	0.884
1.07	1.108	0.666	1.09	0.856	0.889
1.08	1.057	0.630	1.10	0.860	0.894
1.09	1.013	0.597	1.11	0.865	0.897
1.10	0.973	0.568	1.12	0.868	0.900
1.11	0.938	0.543	1.13	0.873	0.905
1.12	0.906	0.519	1.14	0.876	0.909
1.13	0.877	0.498	1.15	0.881	0.913
1.14	0.850	0.479	1.16	0.885	0.917
1.15	0.825	0.461	1.17	0.888	0.920
1.16	0.802	0.445	1.18	0.892	0.924
1.17	0.780	0.430	1.19	0.896	0.928
1.18	0.760	0.415	1.20	0.900	0.932
1.19	0.741	0.402	1.22	0.904	0.935
1.20	0.723	0.390	1.24	0.912	0.943
1.22	0.691	0.366	1.26	0.923	0.952
1.24	0.661	0.346	1.28	0.930	0.957

TABLE 5.3 The Varied-flow Functions for Wide Rectangular Channels (Continued)

$$F(x, y) = \int_0^x \frac{dx}{1 - x^y}$$

	Positive Slope			Negative Slope	
y／x	2.50	3.33	y／x	2.50	3.33
1.26	0.634	0.327	1.30	0.937	0.964
1.28	0.610	0.311	1.32	0.944	0.970
1.30	0.587	0.296	1.34	0.951	0.976
1.32	0.566	0.287	1.36	0.957	0.981
1.34	0.547	0.269	1.38	0.963	0.987
1.36	0.529	0.257	1.40	0.969	0.991
1.38	0.512	0.246	1.42	0.975	0.997
1.40	0.496	0.236	1.44	0.982	1.002
1.42	0.482	0.227	1.46	0.986	1.006
1.44	0.468	0.217	1.48	0.992	1.010
1.46	0.455	0.209	1.50	0.998	1.014
1.48	0.443	0.201	1.55	1.010	1.023
1.50	0.431	0.194	1.60	1.022	1.033
1.55	0.404	0.171	1.65	1.034	1.041
1.60	0.380	0.163	1.70	1.045	1.049
1.65	0.359	0.150	1.75	1.055	1.056
1.70	0.340	0.139	1.80	1.064	1.062
1.75	0.323	0.129	1.85	1.073	1.067
1.80	0.307	0.120	1.90	1.082	1.075
1.85	0.293	0.112	1.95	1.090	1.080
1.90	0.279	0.105	2.00	1.098	1.083
1.95	0.267	0.099	2.10	1.112	1.091
2.00	0.256	0.093	2.20	1.125	1.098
2.10	0.237	0.082	2.3	1.138	1.106
2.20	0.218	0.074	2.4	1.147	1.112
2.3	0.204	0.066	2.5	1.156	1.117
2.4	0.181	0.060	2.6	1.165	1.121
2.5	0.178	0.054	2.7	1.173	1.124
2.6	0.167	0.050	2.8	1.181	1.126
2.7	0.157	0.045	2.9	1.187	1.130
2.8	0.148	0.042	3.0	1.193	1.133
2.9	0.141	0.038	3.5	1.221	1.144
3.0	0.134	0.035	4.0	1.237	1.149
3.5	0.105	0.025	4.5	1.252	1.153
4.0	0.086	0.018	5.0	1.264	1.156
4.5	0.072	0.014	6.0	1.272	1.160
5.0	0.062	0.011	7.0	1.282	1.162
6.0	0.047	0.007	8.0	1.290	1.163
7.0	0.037	0.005	9.0	1.294	1.165
8.0	0.031	0.004	10.0	1.299	1.165
9.0	0.026	0.003			
10.0	0.022	0.002			
20.0	0.008	0.000			

*Values of varied-flow function for positive slopes and negative slopes are interpolated from V. T. Chow, *Open-channel Hydraulics* (New York: McGraw-Hill, 1959), 641–652.

For critical flow in a wide rectangular channel, we also have

$$N_F = \frac{V_c}{\sqrt{gd_c}} = 1.0, \quad \text{or} \quad V_c^2 = gd_c \tag{b}$$

Comparing Eqs. (a) and (b), we have $d_c = \sqrt[3]{\dfrac{q^2}{g}}$

Hence,

$$d_c = \left(\frac{q^2}{g}\right)^{1/3} = \left(\frac{2.5^2}{9.81}\right)^{1/3} = 0.86 \text{ m}$$

To determine the distance between any two sections, we can apply Eq. (5.32), and Table 5.3 as follows:

$$L = x_2 - x_1 = A\left\{(u_2 - u_1) - \left[F\left(u_2, \frac{10}{3}\right) - F\left(u_1, \frac{10}{3}\right)\right]\right.$$
$$\left. + \left[BF\left(v_2, \frac{5}{2}\right) - BF\left(v_1, \frac{5}{2}\right)\right]\right\}$$

As

$$A = \frac{d_n}{S_0} = \frac{1.50}{0.001} = 1500; \qquad B = \left(\frac{d_c}{d_n}\right)^M \frac{J}{N} = \left(\frac{0.86}{1.50}\right)^3 \frac{\left(\dfrac{5}{2}\right)}{\left(\dfrac{10}{3}\right)} = 0.141$$

Water surface profile computations are carried out in the upstream direction with five assumed water depths between $d = 2.00$ m immediately upstream from the dam, and $d = 1.52$ m, which is 1% greater than the normal depth. The computation procedure is shown in Table 5.4.

TABLE 5.4

d	u	v	$F\left(u, \dfrac{10}{3}\right)$	$F\left(v, \dfrac{5}{2}\right)$	$L(m)$	$\Sigma L(m)$
2.00	1.33	1.47	0.277	0.449	−187	187
1.88	1.25	1.35	0.334	0.537	−238	425
1.75	1.17	1.24	0.430	0.661	−325	750
1.64	1.09	1.13	0.597	0.877	−926	1676
1.52	1.01	1.02	1.236	1.595		

If we were only interested in the total distance between the dam and the last section, where $d = 1.52$ m, one step computation alone would be enough.

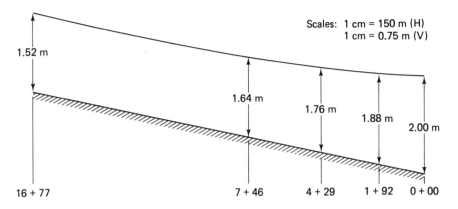

Figure 5.16. Backwater surface profile.

Applying Eq. (5.32), and the corresponding values from the above table, we have

$$L = A\{(u_2 - u_1) - [F(u_2, 3.33) - F(u_1, 3.33)] + B[F(v_2, 2.50)$$
$$- F(v_1, 2.50)]\}$$
$$= 1{,}500\{(1.01 - 1.33) - (1.236 - 0.277) + 0.141(1.595 - 0.449)]\}$$
$$= 1{,}676 \text{ m}$$

Direct Step Method   The direct step method is derived directly from the energy balance between two neighboring sections, 1 and 2, which are separated by a distance sufficiently short that the water surface line can be approximated by a straight line. The energy relation between the two sections may be written as

$$\left(\frac{V_1^2}{2g} + d_1\right) + z = \left(\frac{V_2^2}{2g} + d_2\right) + h_L \qquad (5.33a)$$

where z is the difference in elevation of the channel and h_L is the energy head loss between the two sections. This relationship is also illustrated in Figure 5.17.

Equation (5.33a) may be rewritten in terms defined in Figure 5.17.

$$\left(\frac{V_1^2}{2g} + d_1\right) + S_0 \,\Delta L = \left(\frac{V_2^2}{2g} + d_2\right) + S_e \,\Delta L$$

or

$$E_1 + S_0 \,\Delta L = E_2 + S_e \,\Delta L$$

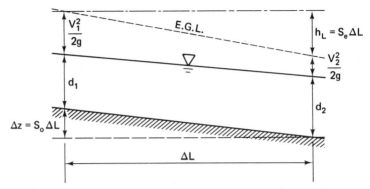

Figure 5.17. Definition for computation of water surface profile.

where E is the specific energy, and

$$\Delta L = \frac{E_1 - E_2}{S_0 - S_e} \tag{5.33b}$$

The computation procedure is to determine the distance ΔL between the section with the known water depth d_1 and the section with an assumed water depth d_2.

The same procedure is applied to the next section in the same direction with d_2 as the known depth. A tabulated computation is particularly recommended as illustrated in Example 5.9. The energy slope, S_e, can be computed by applying Manning's formula

$$S_e = \frac{n^2 V^2}{R_H^{4/3}} \tag{5.34}$$

Example 5.9

Compute the backwater curve on the channel given in Example 5.8 by using the direct step method.

Solution

For a wide rectangular channel, $D \cong R_H \cong d$, and

$$V = \frac{Q}{A} = \frac{q}{d}$$

The conditions given include $S_0 = 0.001$ and $n = 0.025$. From the results of Example 5.8, we have $d_c = 0.86$ m and $d_n = 1.50$ m. From Manning's formula [Eq. (5.34)], $S_e = (n^2 q^2)/d^{10/3}$, which is used to compute the energy slope in Table 5.5.

TABLE 5.5 Backwater Curve Computation—Direct Step Method

(1)	(2)	(3)	(4)	(5)	(6)	(7)	(8)	(9)	(10)	(11)	(12)
d	A	R_H	V	$\dfrac{V^2}{2g}$	E	ΔE	S_e	$\bar{S}_e$	$S_0 - \bar{S}_e$	Δx	L
2.00	—	2.00	1.25	0.080	2.080	—	0.000388	—	—	—	—
1.94	—	1.94	1.29	0.085	2.025	−0.055	0.000429	0.000409	0.000591	−93	93
1.88	—	1.88	1.33	0.090	1.970	−0.055	0.000476	0.000453	0.000547	−101	194
1.82	—	1.82	1.37	0.096	1.916	−0.054	0.000531	0.000504	0.000496	−108	302
1.76	—	1.76	1.42	0.103	1.863	−0.053	0.000593	0.000592	0.000408	−130	432
1.70	—	1.70	1.47	0.110	1.810	−0.053	0.000666	0.000630	0.000370	−143	575
1.64	—	1.64	1.52	0.118	1.758	−0.052	0.000751	0.000709	0.000291	−179	754
1.58	—	1.58	1.58	0.128	1.708	−0.050	0.000850	0.000801	0.000199	−251	1005
1.52	—	1.52	1.64	0.138	1.658	−0.050	0.000967	0.000909	0.000091	−550	1555

Column 1. Depth of flow (in meters) arbitrarily assigned between 2.00 m and the depth, 1.52 m (1 % greater than the normal depth).

Column 2. Water cross-sectional area (in square meters) corresponding to the depth.

Column 3. Hydraulic radius (in meters) corresponding to the depth.

Column 4. Mean velocity (in meters per second) obtained by dividing the discharge by the area in column 2.

Column 5. Velocity head (in meters).

Column 6. Specific energy (in meters) obtained by adding the values from columns 1 and 5.

Column 7. Change in specific energy between two adjacent sections (in meters) obtained by subtracting the E value in column 6 and that of the previous step.

Column 8. Energy slope, or friction slope, obtained from Manning's formula [Eq. (5.34)] by using the values of V in column 4 and R_h from column 3.

Column 9. Average energy slope between the steps, equal to the averaged value of S_e just computed in column 8 and that of the previous step.

Column 10. Difference between the channel bottom slope ($S_0 = 0.001$) and that in column 9.

Column 11. Length (in meters) of the reach computed between the two adjacent sections.

Column 12. Distance from the first section where the water depth is 2.00 m.

PROBLEMS

5.8.1. A wide rectangular channel discharges 1.5 m³/s per unit width. The slope is 0.0009 and the Manning's coefficient is 0.015. At a certain point the water depth is 0.75 m. How far downstream will it be 0.73 m? Use the direct integration method.

5.8.2. A trapezoidal channel with a 5-m bottom width and a $z = 1.0$ side slope discharges 35 m³/s. The slope is 0.004 and it is paved with smooth concrete. Determine the distance between the two sections with depths of 1.69 m and 1.65 m, respectively. Use the direct step method.

5.8.3. A trapezoidal channel paved with smooth concrete has a 2.5-m bottom width and a $z = 1$ side slope. It discharges 18 m³/s on a 0.0004 channel slope. Determine the distance between the two sections with flow depths of 2.16 m and 2.20 m, respectively.

5.8.4. If the channel in Prob. 5.8.3. is discharging 10 m³/s, determine the distance between the section of depth 2.16 m and 2.20 m.

5.8.5. A wide rectangular channel carries 1.6 m³/s per unit width of the channel which has $n = 0.011$ and $S_0 = 0.0016$. If a 5-m height dam is placed across the channel, determine the water surface profile. Use the direct integration method with 0.30 m depth increments.

5.8.6. The downstream face of the dam in Prob. 5.8.5 has a slope of 1/10 until it reaches the channel bottom again (see Figure 5.18). Determine the water surface profile over the downstream face of the dam. Use the direct integration method with depth increments of your choice.

Figure 5.18

5.8.7. Determine the water surface profile in the downstream channel immediately after the dam in Prob. 5.8.6.

5.8.8. Compute the water surface profile if a dam is placed across the channel of Prob. 5.8.2 to raise the water level to a height of 3.4 m above the channel bottom immediately behind the dam.

5.8.9. A long trapezoidal channel with a roughness coefficient of 0.015, bottom width measuring 3.6 m, and $z = 2.0$ discharges 44 m³/s. If the channel slope is 0.0001, compute the water surface profile upstream from a dam which is 5 m in height.

5.9 TRANSITIONAL PHENOMENA IN SUPERCRITICAL FLOWS

Physically, the basic difference between supercritical and subcritical flows is the way surface waves are transmitted in a channel. In supercritical flow a small surface wave created by any disturbance is washed downstream because the velocity of flow in the channel is greater than the speed of the wave. In subcritical flow the velocity of the flow is less than the speed of the wave; the wave will propagate upstream as well as downstream. The patterns of surface wave propagation in stationary water ($V = 0$), subcritical flow ($V < \sqrt{gD}$), critical flow ($V = \sqrt{gD}$), and supercritical flow ($V > \sqrt{dD}$) are schematically illustrated in Figure 5.19.

In Section 5.4 we said that the speed of a small surface wave traveling on water is equal to $\sqrt{gD}$, where D is the local water depth and g is the gravitational acceleration. If the velocity of water is V, then the absolute velocity of a surface wave traveling in the upstream direction is

$$V - \sqrt{gD}$$

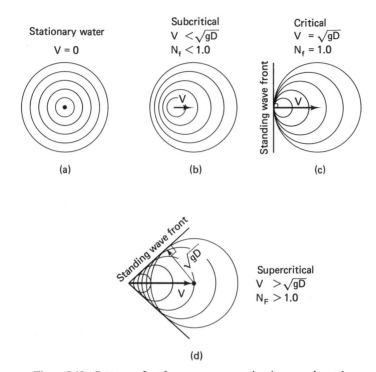

Figure 5.19. Patterns of surface wave propagation in open channel.

In the case of critical flow, $V = \sqrt{gD}$, a standing wave front is created, as shown in Figure 5.19(c). The wave front in critical flow is normal to the direction of the flow.

In supercritical flows the speed of water flow V exceeds $\sqrt{gD}$. The standing wave front forms a wave angle β, as shown in Figure 5.19(d). The angle β obviously decreases as the flow speed increases. If a point P is taken on the wave front, it is easily seen that the location of P after a unit time can be determined by the longitudinal displacement O–O' and the radial displacement O'–P. During a unit time period the center of disturbance O is displaced by a distance equal to the water velocity V (represented by O–O'); and the point P is carried away from the center by a radial distance equal to the wave speed $\sqrt{gD}$ (represented by O'–P), as shown in Figure 5.20. From the geometry, the magnitude of the wave angle can be determined as

$$\sin \beta = \frac{\sqrt{gD}}{V} = \frac{1}{\dfrac{V}{\sqrt{gD}}} = \frac{1}{N_F} \qquad (5.35)$$

This relationship is analogous to the sound wave phenomenon in supersonic gas flows. The oblique standing wave front in supercritical open channel

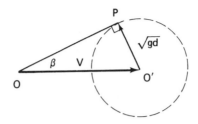

Figure 5.20. Wave angle in supercritical flow.

flow is analogous to the shock wave created at the leading front of a super-sonic airplane.

Deflection of a channel wall creates a definite disturbance to the super-critical flow. The standing wave generated by a channel wall deflection may generally be classified into two categories: (1) a positive deflection if the wall deflects into the flow and (2) a negative deflection if the wall deflects away from the flow. The deflection wave front propagates across the channel, as shown in Figure 5.21. Water passing through a deflection wave front increases its water depth if the deflection is positive and decreases it if the deflection is negative. Given the angle of wall deflection, θ, the wave angle β may be determined by considering the following factors.

Assuming flow depth d_1 and velocity V_1 for the approaching flow and d_2 and V_2 for that after the flow transverses the wave front, we may resolve the velocities V_1 and V_2 into components parallel and perpendicular to the wave front, as shown in Figure 5.22(a). The components parallel to the wave front must be equal to each other since no obvious shear should exist at the front. Figure 5.22(a) shows this relationship as

$$V_1 \cos \beta = V_2 \cos (\beta - \theta) \tag{5.36}$$

According to the continuity equation, we may write another relationship between the upstream section and the downstream section of the wave front as

$$d_1 V_1 \sin \beta = d_2 V_2 \sin (\beta - \theta) \tag{5.37}$$

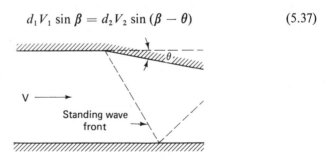

Figure 5.21. Deflection standing wave in supercritical channel.

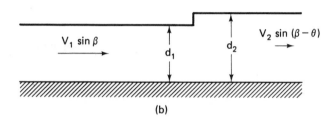

Figure 5.22. Description of a standing wave front at wall deflection. (a) Top view. (b) Section view.

In the meantime, the momentum equation per unit width of the channel upstream and downstream from the wave front requires that

$$F_1 + pq(V_1 \sin \beta) = F_2 + pqV_2 \sin (\beta - \theta)$$

Substituting

$$F_1 = \frac{\gamma d_1^2}{2}, \qquad F_2 = \frac{\gamma d_2^2}{2}, \qquad q = d_1 V_1 \sin \beta = d_2 V_2 \sin (\beta - \theta)$$

into the above equation, we have

$$\frac{\gamma d_1^2}{2} + p d_1 V_1^2 \sin^2 \beta = \frac{\gamma d_2^2}{2} + p d_2 V_2^2 \sin^2 (\beta - \theta) \qquad (5.38)$$

From Eq. (5.37) we have

$$V_2 = \frac{d_1 V_1 \sin \beta}{d_2 \sin (\beta - \theta)}$$

This relationship may be substituted into Eq. (5.38) and simplified for the solution of the wave angle β,

$$\sin \beta = \frac{1}{N_{F_1}}\sqrt{\left(\frac{1}{2}\right)\left(\frac{d_2}{d_1}\right)\left(1 + \frac{d_2}{d_1}\right)} \qquad (5.39)$$

where $N_{F_1} = V_1/\sqrt{gD_1}$ is the Froude number of the approaching flow. The ratio d_2/d_1 is greater than unity for a positive deflection wave front and less than unity for a negative deflection wave front. For very small disturbances, the ratio d_2/d_1 is near unity and Eq. (5.39) simplifies to Eq. (5.35).

From the continuity equation (5.37) the depth ratio may also be expressed as

$$\frac{d_2}{d_1} = \frac{V_1 \sin \beta}{V_2 \sin (\beta - \theta)} \qquad (5.40)$$

Substituting the above expression into Eq. (5.39) and rearranging the terms, we have

$$\frac{\tan \beta}{\tan (\beta - \theta)} = \left(\frac{1}{2}\right)[-1 + \sqrt{1 + (8 \sin^2 \beta)N_{F_1}^2}] \qquad (5.41)$$

The above transcendental equation expresses the relationship between the wall deflection angle θ and the wave angle β. It may be solved for β if the condition of the approaching flow, N_{F_1}, is given.

The general relations among the deflection angle, θ, the Froude number of the approaching flow, N_{F_1}, the depth ratio, d_2/d_1, and the wave angle, β, have been presented by A. T. Ippen as shown in Figure 5.23.

Example 5.10

A rectangular channel 3.3 m wide carries a discharge of 4.95 m³/s. The flow encounters a positive wall deflection of $\theta = 5°$. The depth upstream from the wave is 0.5 m. Determine the standing wave angle and the water depth downstream of the wave front.

Solution

Substituting the given conditions $\theta = 5°$, $d_1 = 0.5$ m, and

$$V_1 = \frac{Q}{Bd_1} = \frac{4.95}{3.3 \cdot 0.5} = 3 \text{ m/s}$$

into Eq. (5.41) gives

$$\frac{\tan \beta}{\tan (\beta - 5°)} = \frac{1}{2}\left[-1 + \sqrt{1 + 8(\sin^2 \beta)\left(\frac{3^2}{g \cdot 0.5}\right)}\right]$$

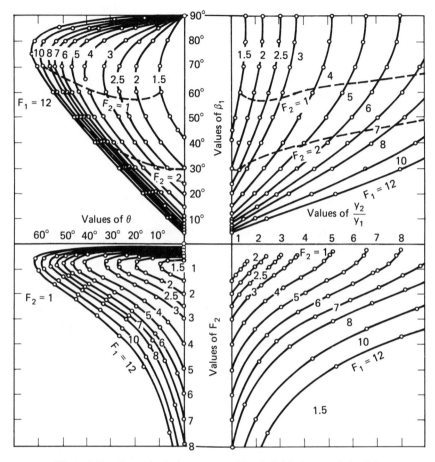

Figure 5.23. General relations among N_{b1}, θ, d_2/d_1, N_{F_2}, and β. (After A. T. Ippen*).

Solution may be obtained by trial. We find

$$\beta = 58°27'(58.45°)$$

The downstream depth can then be determined by using Eq. (5.39). Multiplying both sides of the equation and squaring give

$$2N_{F_1}^2 \sin^2 \beta = 2\left(\frac{V_1^2}{gd_1}\right)\sin^2 \beta = \frac{d_2}{d_1}\left(1 + \frac{d_2}{d_1}\right) = 2.665$$

*A. T. Ippen, "Mechanics of supercritical flow, 1st paper of high-velocity flow in open channels: A symposium," *Trans. ASCE, 116*, (1956), 678–694.

Hence,

$$\frac{d_2}{d_1} = \frac{-1 \pm \sqrt{1 + 10.66}}{2} = -0.5 \pm 1.707$$

Here only the positive value is taken since a negative water depth has no physical meaning. Therefore,

$$d_2 = 1.21 d_1 = 0.605 \text{ m}$$

PROBLEMS

5.9.1. Repeat Example 5.10 if the initial depth and the wall deflection were, respectively, 0.45 m and 8°.

5.9.2. A wide rectangular channel discharging 3 m³/s per unit width encounters a positive deflection of 15°. If the wave angle is 45°, determine the upstream and downstream depths.

5.9.3. A 5-m rectangular channel which carries 15 m³/s encounters a wall deflection. The water depths upstream and downstream are measured, respectively, as 0.6 m and 0.7 m. Determine the wave angle and the angle of wall deflection.

5.9.4. Water flows in an open channel at 4 m/s with a depth of 0.4 m. Determine the wave angle when the channel encounters a wall deflection of 10°.

5.10 WAVE INTERFERENCE AND DESIGN OF SUPERCRITICAL TRANSITIONS

Deflection wave fronts may be generated from more than one location in a supercritical flow structure. The standing wave fronts, positive or negative or both, meet in the flow field and form new regions of deflections. The complex surface phenomena caused by wave interference may be determined by applying the wave equations stated previously.

Figure 5.24. illustrates schematically the wave interference in a super-critical flow channel. Clearly, a standing wave pattern of this kind is not satisfactory for channel operation. With a properly chosen deflection angle, a transition may be designed to minimize the wave interference and produce a uniform flow condition downstream from the transition.

The most effective design of a *contraction in supercritical channel* is a straight-wall contraction, as shown in Figure 5.25. If the deflection angle, θ, and the transition length, L, are properly chosen, the wave front reflected from the first meeting point B of the two initial waves AB and CB will arrive at the side walls exactly at the end of the transition, E and E'. This requirement is met if the following geometric condition is satisfied:

$$L = \frac{B_1}{2 \tan \beta_1} + \frac{B_2}{2 \tan (\beta_2 - \theta)} \qquad (5.42)$$

(a)

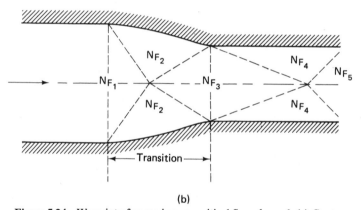

(b)

Figure 5.24. Wave interference in supercritical flow channel. (a) Contraction. (b) Bend.

The wave angle β_1 and the transition Froude number, N_{F_2}, are determined by the approaching flow conditions

$$N_{F_1} = \frac{V_1}{\sqrt{gd_1}}$$

and the angle of wall deflection, θ. The wave angle β_2 in turn depends on θ and N_{f_2}. On the other hand, the simple geometry of the transition also requires that

$$L = \frac{B_1 - B_2}{2 \tan \theta} \tag{5.43}$$

The proper value of θ may thus be determined by satisfying both Eqs. (5.42) and (5.43) or by solving the value of θ from the following combined

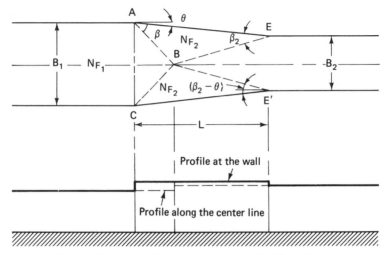

Figure 5.25. Contraction design in supercritical flow channel.

relationship:

$$\frac{B_1 - B_2}{2 \tan \theta} = \frac{B_1}{2 \tan \beta_1} + \frac{B_2}{2 \tan (\beta_2 - \theta)} \qquad (5.44)$$

An improperly designed *expansion in a supercritical channel* usually encounters serious wave problems at the downstream end where the expansion terminates. The connection of the expansion to the walls of the exit channel sets up a positive wave front (Figure 5.26) and may become a serious problem in the downstream operation.

Although it is possible to design an expansion based strictly on the theory of wave mechanics to avoid the downstream wave interference, the theoretical design was found usually to require an impractically long transition. Using a series of analytical and experimental studies as a basis, H. Rouse and his colleagues* developed a family of generalized expansion wall curves that

Figure 5.26. Positive waves in supercritical exit channel.

*H. Rouse, B. V. Bhoota, and E. Y. Hsu, "Design of Channel Expansions," *Trans. ASCE.*, *116* (1951), 347.

give the geometry of the expansion walls covering a wide range of width ratios, B_1/B_2, for a variety of approaching flow conditions. This family of curves has been successfully used in the design of expansions which give an approximately rectilinear flow at the outlet and allow the downstream channel to be free from gross surface disturbances. The expansion curves are shown in Figure 5.27.

In a bend of a supercritical flow channel, abrupt angle changes such as the one shown in Figure 5.24(a) should be avoided in order to minimize the

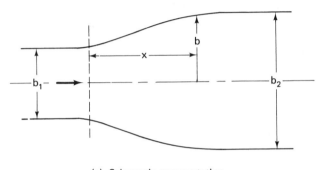

(a) Schematic representation
of a channel expansion

(b) Expansion curves

Figure 5.27. Generalized curves for supercritical channel expansion (after Rouse, Bhoota, and Hsu, 1951).

standing wave interference. A curved channel bend with a simple smooth curve causes super-elevation of water surface near the outer wall at the bend. The standing wave interference pattern, with alternative peaks and troughs, is superimposed on the elevated water surface in a rapid flow, as shown in Figure 5.28.

Two different methods may be used to eliminate the excessive rise of water surface elevation at a supercritical channel bend. The first method is known as *water surface banking* and is accomplished by physically dropping the inner channel bottom elevation at the bend and leaving the outer bottom at the same level. In this arrangement, a transitional length must be provided both before and after the main curve in consideration of the secondary current developed in the bend. The recommended transitional length for a rectangular channel is

$$L = 15\left(\frac{V^2B}{gR}\right) \tag{5.45}$$

where V is the mean velocity, B is the channel width, and R is the centerline radius at the bend.

The second method uses a compound curve for the bend, as shown in Figure 5.29. The compound curve usually includes a series of three curves: a main curve and two *transition curves*, each at one end of the main curve.

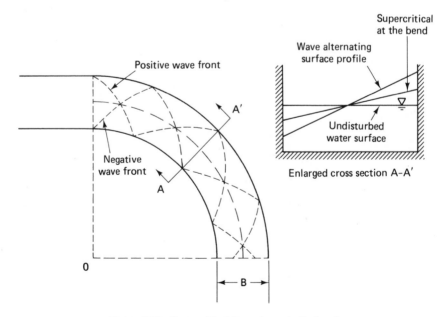

Figure 5.28. Supercritical flow channel at a bend.

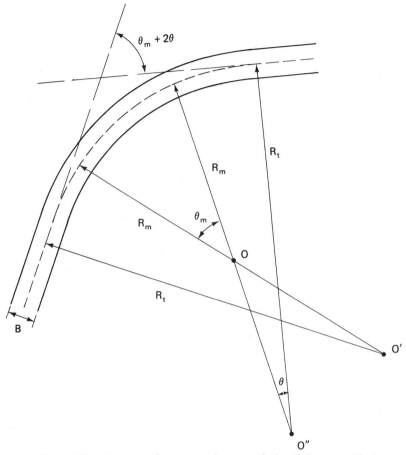

Figure 5.29. Geometry of a compound curve at the bend of a supercritical channel.

With different radii of curvature and central angle from the main curve, the transition curves produce a disturbance wave pattern that offsets the wave pattern created by the main curve at a fixed phase angle. The geometry of the transition curves can be selected so that a maximum depth occurs just at the point where the disturbance due to the main curve would begin. This way, the negative disturbance from the beginning of the inner transition curve reaches the outer wall just in time to cancel the positive disturbance that would be generated there, and, therefore, no disturbance is generated.

The transition curves should be made with radius equal to twice that of the main curve,

$$R_t = 2R_m \tag{5.46}$$

and the central angle of the transition curves may be approximated by

$$\theta = \tan^{-1}\left[\frac{\left(\dfrac{B}{\tan \beta_m}\right)}{R_t + 0.5B}\right] \tag{5.47}$$

where β_m is the wave angle of the main curve when subjected to the undisturbed flow conditions.

Example 5.11

A rectangular channel carrying a discharge of 20 m³/s is required to contract from a width of 3.5 m to a width of 2.8 m. If the channel slope is 0.01 and Manning's $n = 0.011$, design the most effective transition for the contraction so that the wave interference can be prevented downstream.

Solution

If a value is assumed for θ, the length of the transition can be computed by Eq. (5.43):

$$L = \frac{B_1 - B_2}{2 \tan \theta}$$

and the wave angle by Eq. (5.41):

$$\frac{\tan \beta_1}{\tan (\beta_1 - \theta)} = \frac{1}{2}[-1 + \sqrt{1 + 8(\sin \beta_1 N_{F_1})^2}]$$

Figure 5.23 gives the value for a first approximation. The velocity and the Froude number in the transition region can be determined through Eqs. (5.36) and (5.40):

$$V_2 = V_1 \frac{\cos \beta_1}{\cos (\beta_1 - \theta)}$$

and

$$d_2 = d_1 \frac{V_1}{V_2}\frac{\sin \beta_1}{\sin (\beta_1 - \theta)} = d_1 \frac{\tan \beta_1}{\tan (\beta_1 - \theta)}$$

so that

$$N_{F_2} = \frac{V_2}{\sqrt{gd_2}} = \frac{V_2}{\sqrt{gd_1 \dfrac{\tan \beta_1}{\tan (\beta_1 - \theta)}}}$$

The second wave angle is determined similarly.

$$\frac{\tan \beta_2}{\tan (\beta_2 - \theta)} = \frac{1}{2}[1 + \sqrt{1 + 8(\sin \beta_2 N_{F_2})^2}]$$

Eq. (5.42) is now used to check the length:

$$L = \frac{B_1}{2 \tan \beta_1} + \frac{B_2}{2 \tan (\beta_2 - \theta)}$$

A new wall deflection, θ, is chosen and the procedure repeated until the length checks fairly well. Table 5.6 was developed to follow this procedure, where $A = 1/2[-1 + \sqrt{1 + 8(\sin \beta \, N_F)^2}]$

TABLE 5.6

θ	$L(m)$	β_1	$\frac{\tan \beta_1}{\tan(\beta_1-\theta)}$	A_1	V_2	N_{f_2}	β_2	$\frac{\tan \beta_2}{\tan(\beta_2-\theta)}$	A_2	$L(m)$
5°	4.00	25°	1.2812	1.2300						
		26°	1.2706	1.2892						
		25.75°	1.2731	1.2744	7.27	2.365	31.00°	1.2319	1.2932	
							30.00°	1.2381	1.2455	
							29.85°	1.2391	1.2382	6.65
2.7°	7.42	23°	1.1475	1.1108						
		23.5°	1.1447	1.1407						
		23.56°	1.1443	1.1442	7.406	2.541	27.65°	1.1261	1.1845	
							25.70°	1.1338	1.1366	
							25.65°	1.1340	1.1339	7.32
2.78°	7.21	23.64°	1.1487	1.1490	7.402	2.534	25.8°	1.1378	1.1379	7.29
2.74°	7.31	23.6°	1.1465	1.1466	7.404	2.537	25.73°	1.1359	1.1359	7.31

The wall deflection should be 2.74° and the transition length is 7.31 m.

PROBLEMS

5.10.1. Design a proper expansion for the channel in Example 5.11 in which the width will change from 2.8 m to 5.6 m.

5.10.2. A rectangular channel 4-m wide carries a discharge of 18 m³/s at a normal depth of 0.6 m. The channel must be contracted to a width of 3 m. Design a transition that will prevent wave interference in the downstream channel.

5.10.3. A 4-m wide supercritical channel has a rectangular cross section and it carries 40 m³/s discharge. The flow approaches a transition with the depth of 0.6 m. Design the transition to narrow the width of the channel by 50%. Determine the water velocity and the depth at the end of the transition.

6

HYDRAULICS OF
WELLS AND SEEPAGE

Groundwater occurs in permeable, water-bearing geologic formations known as *aquifers*. There are basically two types of aquifer: The *confined aquifer* is a relatively high-permeable, water-bearing formation (e.g., sand or gravels) confined below a layer of very low permeability (e.g., layer of clay). The *unconfined aquifer* is a water-bearing formation with a free *water table*, below which the soil is saturated and above which there is air under atmospheric pressure. Movement of ground water occurs under hydraulic gradient or gravitational slope in the same way as water movement occurs in pipes or open channels.

Pressure gradient in an aquifer may occur naturally in courses (water table gradient) or it may occur as a result of artificial means (such as well pumps). Figure 6.1 schematically shows several examples of ground water occurrence in both confined and unconfined aquifer formations.

The pressure level (head) in a confined aquifer is represented by the piezometric surface which is usually originated from a distant source, such as the water table of a recharging area. An artesian spring is formed if the impermeable confining stratum is perforated at a location where the ground surface falls below the piezometric surface. It becomes an artesian well if the ground surface raises above the piezometric surface. The water table of an unconfined aquifer is usually unrelated to the piezometric surface of a con-

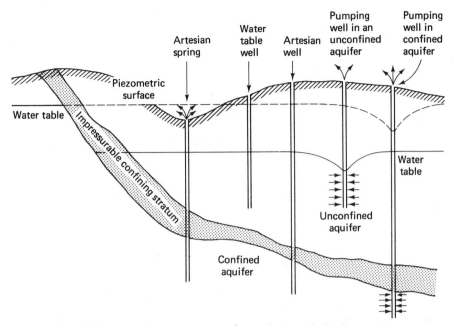

Figure 6.1. Ground water occurrence in confined and unconfined aquifer.

fined aquifer in the same region, as shown in Figure 6.1. This is because the confined and unconfined aquifers are hydraulically separated by the impermeable confining stratum.

The capacity of a ground water reservoir in a particular geological formation depends strictly on the percentage of void space in the formation and on how the interstitial spaces are connected. Figure 6.2 shows schematically several types of rock formation and their relation to the interstitial space.

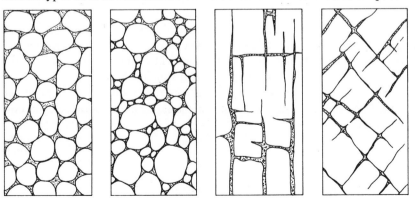

Figure 6.2. Examples of rock texture and interstices.

The percentage ratio of void space to the total volume of the formation is known as the *porosity* of the formation, defined as

$$\alpha = \frac{100\omega}{\text{Vol}} \tag{6.1}$$

where ω is the volume of the void space and Vol is the total volume of the formation.

TABLE 6.1 Porosity Range in Common Water-Bearing Formations

Material	$\alpha(\%)$
Clay	45–55
Silt	40–50
Medium to coarse mixed sand	35–40
Fine to medium mixed sand	30–35
Uniform sand	30–40
Gravel	30–40
Gravel and sand mixed	20–35
Sandstone	10–20
Limestone	1–10

6.1 MOVEMENT OF GROUND WATER

Movement of ground water is proportional to the hydraulic gradient in the direction of the flow. The *apparent velocity* of ground water movement in a porous medium is governed by *Darcy's law**

$$V = K\frac{dh}{dL} \tag{6.2}$$

where dh/dL is the hydraulic gradient in the direction of the flow path (dL) and K is a proportionality constant known as the *coefficient of permeability*. The apparent velocity is defined by the quotient of discharge divided by the cross-sectional area of the medium through which it flows.

The coefficient of permeability depends not only on the medium of the formation but also on the properties of the fluid. By dimensional analysis, we may write

$$K = \frac{Cd^2\gamma}{\mu} \tag{6.3}$$

where Cd^2 is a property of the medium formation only, γ is the specific gravity, and μ is the viscosity of the fluid, respectively. The constant C repre-

---

*H. Darcy, *Les Fontaines Publiques de la Ville de Dijon* (Paris: V. Dalmont, 1856).

sents the various properties of the medium formation that affect the flow other than d, which is a representative dimension proportional to the size of the interstitial space in the medium formation. The coefficient of permeability can be effectively determined by laboratory experiments or field tests applying Darcy's equation [Eq. (6.2)], as will be discussed later.

Physically, the apparent velocity is the speed at which the bulk of ground water migrates between two stations ΔL apart in the porous medium.

$$V = K \frac{\Delta L}{\Delta t}$$

This is not the actual speed that an individual water particle travels in the pores. The actual distance that water particles travel between any two stations ΔL apart in the porous medium cannot be a straight line; hence it must be longer than ΔL. In this chapter the movement of ground water is treated from a hydraulic engineer's point of view. We are only interested in the movement of water in quantity. The fluid mechanical aspect of water particle movement in pores will not be included in this discussion.

For an area A perpendicular to the direction of the water movement in an aquifer, the discharge may be expressed as

$$Q = AV = KA \frac{dh}{dL} \qquad (6.4)$$

Laboratory measurement of the coefficient of permeability can be demonstrated by the following example.

Example 6.1

A small sample of an aquifer is packed in a test cylinder (see Figure 6.3) to form a column 30 cm long and 4 cm in diameter. At the outlet of the test cylinder, 21.3 cm³ of water is collected in 2 min. During the testing period a constant piezometric head difference of $\Delta h = 14.1$ cm is observed. Determine the coefficient of permeability of the aquifer sample.

Solution

The cross-sectional area of the sample volume is

$$A = \frac{\pi}{4}(4)^2 = 12.56 \text{ cm}^2$$

The hydraulic gradient is equal to the change of piezometric head per unit length of the aquifer measured in the direction of the flow:

$$\frac{dh}{dL} = \frac{14.1}{30} = 0.47$$

Figure 6.3. Laboratory determination of coefficient of permeability.

The discharge rate

$$Q = 21.3 \text{ cm}^3/2 \text{ min} = 10.65 \text{ cm}^3/\text{min}$$

Applying Darcy's law [Eq. (6.4)], we may calculate

$$K = \frac{Q}{A} \frac{1}{\dfrac{dh}{dL}} = \frac{10.65}{12.56} \cdot \frac{1}{0.47} = 1.80 \text{ cm}^3/\text{min}/\text{cm}^2$$

Based on Darcy's law, several different types of permeameters have been developed for laboratory measurements of permeability in small samples of aquifers.* Although laboratory testing is conducted under controlled conditions, such measurements may not represent the field permeability. When unconsolidated samples are taken from the field and repacked in laboratory permeameters, the texture, porosity, grain orientation, and packing may be markedly disturbed and changed; consequently, permeabilities are modified. For better reliability, field permeability of aquifer can be determined by well pumping tests in the field. This method will be discussed in Section 6.3.

---

*L. K. Wenzel, *Methods for Determining Permeability of Water-bearing Materials with Special Reference to Discharging-Well methods*, U.S.G.S. Water Supply Paper #887 (1942).

PROBLEMS

6.1.1. In a laboratory experiment a sand sample is packed into a cylindrical sample space 4 cm in diameter and 20 cm long. Under a steady head of 24 cm, 100 cm³ of water is collected in 5 min. What is the apparent velocity of water flowing through the sample? Determine the coefficient of permeability of the sample.

6.1.2. Field measurement of a uniform aquifer indicated that the aquifer has a cross-sectional area of 1000 m². Two wells 25 m apart are drilled into the ground along the direction of the flow. It is observed that the water surface elevation difference between the two wells is 0.35 cm. A salt tracer is dissolved in the upstream well and is picked up in the downstream well 42 h later. Determine the coefficient of permeability of the aquifer.

6.1.3. The experiment in Prob. 6.1.1. was done at room temperature (20°C). Apply Eq. (6.3) and determine the discharge in 5 min if the same experiment had been performed at a temperature of 5°C. What is the coefficient of the sample at this temperature?

6.1.4. A sand filter as shown in Figure 6.4 is 1 m deep and has a surface area of 4 m². If the permeability is 0.65 cm/s, determine the discharge through the filter for a differential head of 0.8 m.

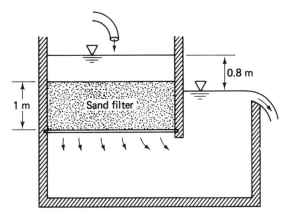

Figure 6.4

6.2 RADIAL FLOW TO A WELL

Removal of water from the aquifer surrounding a well by pumping, results in radial flow to the well, because the pumping action lowers the water table (or the piezometric surface) at the well and forms a region of pressure depression that surrounds the well. At any given distance from the well the *drawdown* of the water table (or piezometric surface) is defined by the vertical distance measured from the original to the lowered water table (or piezometric surface). Figure 6.5(a) shows the drawdown curve of the water table in an unconfined aquifer; Figure 6.5(b) shows the drawdown curve of the piezometric surface in a confined aquifer.

In a uniform aquifer formation the axisymmetric drawdown curve describes a conic-shaped geometry commonly known as the *cone of depression*. The outer limit of the cone of depression defines the *area of influence* of the well.

Darcy's law may be directly applied to derive the *radial flow equation* which relates the discharge to the drawdown of the water table at a well in the unconfined aquifer after a steady state of equilibrium is reached. By using plane polar coordinates with the well as the origin, the discharge to the well at any distance r from the well may be expressed as

$$Q = AV = 2\pi rh\left(K\frac{dh}{dr}\right) \tag{6.5}$$

Integrating between the limits of the well (where $r = r_w$, $h = h_w$) and the undisturbed water table (where $r = r_0$, $h = h_0$) yields

$$Q = \pi K\frac{h_0^2 - h_w^2}{\left(\ln\frac{r_0}{r_w}\right)} \tag{6.6}$$

The selection of the radius of influence, r_0, is arbitrary. The variation of Q is rather small for the wide range of r_0, because the influence on the well by the water table at far distances is small. In practice, approximate values of r_0 may be taken between 100 and 500 m, depending on the nature of the aquifer and the operation of the pump.

In confined aquifers, water delivery to a well is confined to the depth of well penetration into the aquifer, b. Applying Darcy's law, we find that the discharge flowing through the cylindrical boundary with radius r from the center of the well equals

$$Q = AV = 2\pi rb\left(K\frac{dh}{dr}\right) \tag{6.7}$$

Integrating between the boundary conditions at the well ($r = r_w$, $h = h_w$) and at the radius of influence ($r = r_0$, $h = h_0$) yields

$$Q = 2\pi Kb\frac{h_0 - h_w}{\ln\left(\frac{r_0}{r_w}\right)} \tag{6.8}$$

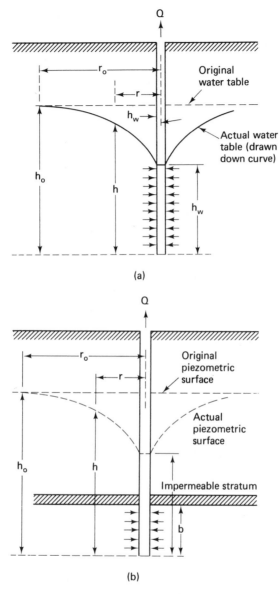

(a)

(b)

Figure 6.5. Radial flow to a pumping well from (a) an unconfined aquifer and (b) a confined aquifer.

The above equations apply only to steady flow where the discharge remains a constant value at any distance from the well. A more general equation for the discharge may be written for any distance, r:

$$Q = 2\pi Kb \frac{h - h_w}{\ln\left(\dfrac{r}{r_w}\right)} \qquad (6.9)$$

Eliminating Q between Eqs. (6.8) and (6.9) results in

$$h - h_w = (h_0 - h_w) \frac{\ln \left(\dfrac{r}{r_w}\right)}{\ln \left(\dfrac{r_0}{r_w}\right)} \qquad (6.10)$$

which shows that the piezometric head varies linearly with the logarithm of the distance from the well, regardless of the rate of discharge.

6.3 FIELD DETERMINATION OF PERMEABILITY COEFFICIENT

Laboratory tests of soil permeability (Example 6.1) are performed on small samples of soil. Their value in solving engineering problems depends on how well they represent the masses of materials in the field. When used with consideration of field conditions and careful handling of the samples, the laboratory test methods can be very valuable. Nevertheless, important hydraulic engineering projects often require field tests which measure the undisturbed soil permeabilities *in situ*.

EQUILIBRIUM TEST

The aquifer permeability coefficient, K, can be effectively measured in the field by well pumping tests.

In addition to the pumped well, the pumping test requires two observation wells to penetrate the aquifer. The observation wells are located at two arbitrary distances, r_1 and r_2, from the pumped well, as schematically repre-

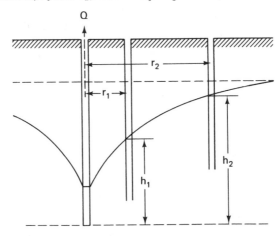

Figure 6.6. Field determination of permeability coefficient in unconfined aquifers.

sented in Figure 6.6. After pumping the well at a constant discharge Q for a long period, the water levels in the observation wells, h_1 and h_2, will reach final equilibrium values. The equilibrium water levels in the observation wells are measured to calculate the aquifer permeability coefficient.

For unconfined aquifers, the coefficient can be calculated by integrating Eq. (6.6) between the limits of the two observation wells:

$$K = \frac{Q}{\pi(h_2^2 - h_1^2)} \ln\left(\frac{r_2}{r_1}\right) \tag{6.11}$$

For confined aquifers, the coefficient can be calculated by rearranging Eq. (6.8) as follows:

$$K = \frac{Q}{2\pi b(h_2 - h_1)} \ln\left(\frac{r_2}{r_1}\right) \tag{6.12}$$

where b is the thickness of the confined aquifer, as shown in Figure 6.7.

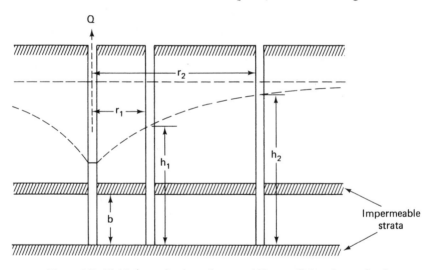

Figure 6.7. Field determination of permeability coefficient in confined aquifers.

The derivation of Eqs. 6.11 and 6.12 assumes that the aquifer is homogeneous and isotropic and that it extends over an area larger than the radius of influence can reach. In practice, however, the equations have been widely used to provide good estimations of permeability coefficients for a variety of aquifer conditions, in spite of these restrictive conditions.

Table 6.2 gives an indication of the numerical values for the coefficient of permeability for some natural soil formations.

TABLE 6.2 Magnitude of Coefficient of Permeability of Some Natural Soil Formations

Soils	K (m/s)
Clays	$< 10^{-9}$
Sandy clays	$10^{-9}-10^{-8}$
Peat	$10^{-9}-10^{-7}$
Silt	$10^{-8}-10^{-7}$
Very fine sands	$10^{-6}-10^{-5}$
Fine sands	$10^{-5}-10^{-4}$
Coarse sands	$10^{-4}-10^{-2}$
Sand with gravels	$10^{-3}-10^{-2}$
Gravels	$> 10^{-2}$

Example 6.2

A well 20 cm in diameter penetrates 30 m deep into the undisturbed water table of an unconfined aquifer. After a long period of pumping at the constant rate of 0.1 m³/s, the drawdown at 20 m and 50 m distance from the well are observed to be 4 m and 2.5 m, respectively. Determine the coefficient of permeability of the aquifer. What is the drawdown at the pumped well?

Solution

Conditions given are $Q = 0.1$ m³/s, $r_1 = 20$ m, $r_2 = 50$ m; hence, $h_1 = 30.0$ m $- 4$ m $= 26$ m, and $h_2 = 30.0$ m $- 2.5$ m $= 27.5$ m, in reference to Figures 6.5 and 6.6. Substituting these values into Eq. (6.11), we have

$$K = \frac{0.1}{\pi(27.5^2 - 26^2)} \ln\left(\frac{50}{20}\right) = \frac{0.00114}{\pi} = 0.000363 \text{ m/s}$$

The drawdown at the pumped well can be calcuated by using the same equation with the calculated value of the coefficient of permeability and the well diameter.

At the well

$$r = r_0 = \frac{d_0}{2} = \frac{0.2}{2} = 0.1 \text{ m}$$

and

$$h = h_0$$

Therefore,

$$\frac{0.00114}{\pi} = \frac{Q}{\pi(h_1^2 - h_0^2)} \ln\left(\frac{r_1}{r_0}\right) = \frac{0.1}{\pi(26^2 - h_0^2)} \ln\left(\frac{20}{0.1}\right)$$

From which we have

$$h_0 = 14.53 \text{ m}$$

The drawdown at the well is $30 - 14.53 = 15.47$ m.

PROBLEMS

6.3.1. A confined aquifer of uniform thickness (18 m) is completely penetrated by a pumping well. After a long period of pumping at the constant rate of 0.3 m³/s, the water elevations in the observation wells ($r_1 = 20$ m, $r_2 = 65$ m) are stabilized. The drawdown measured at the observation wells are, respectively, 16.25 m and 3.42 m. Determine the coefficient of permeability.

6.3.2. A sample of an aquifer is taken from a circular island in a lake to a laboratory where the coefficient of permeability is determined to be 0.000142 m/s. The island (see Figure 6.8) is approximately 800 m in diameter, and a 30-cm diameter well is installed at the center of the island. Determine the minimum depth that the well must penetrate below the surface elevation of the lake in order to yield a constant discharge of 0.2 m³/s.

Figure 6.8

6.3.3. Obtain the free surface of the water table for the well in Figure 6.9. If the well yields 1.5 m³/h, what is the permeability coefficient?

Figure 6.9

6.3.4. If water is taken from the well in Prob. 6.3.3. at a rate of 1.85 m³/h, what would be the water surface elevation in the well as measured with respect to the impervious floor?

NONEQUILIBRIUM TEST

Field determination of permeability K using the equilibrium theory assumed that a new well must be pumped at a constant flow rate until an equilibrium condition is established. In reality, however, many years may be required for true equilibrium to develop in many confined aquifers. It is usually not possible to reach even approximate equilibrium when a new well is pumped for a reasonable time.

C.V. Theis* first presented an analysis of well flow which took into consideration the effect of time and the storage characteristic of the confined aquifer. The Theis equation is

$$h_0 - h = \frac{Q}{4\pi T} \int_u^\infty \frac{e^{-u}}{u}\, du \qquad (6.13)$$

where $h_0 - h$ is the drawdown in an observation well at a distance r from the pumping well and u is a dimensionless parameter given by

$$u = \frac{r^2 S_c}{4Tt} \qquad (6.14)$$

In Eq. (6.14) t is the time since the pumping began, and S_c is the *storage constant* of the aquifer, a dimensionless number defined as the water yield from a column of aquifer of unit area by lowering the water table or piezometric surface by one unit height. T is the *transmissibility* of the aquifer, defined as

$$T = \text{(aquifer thickness)(permeability)} = bK$$

The integral in Eq. (6.13) is commonly written as $W(u)$, known as the well function of u. It is not directly integratable but may be evaluated by the series

$$W(u) = -0.5772 - \ln u + u - \frac{u^2}{2\cdot 2!} + \frac{u^3}{3\cdot 3!} \cdots \qquad (6.15)$$

For a reasonably large value of t, and a small value of r, u can be made small enough that the terms following $\ln u$ become small and may be neglected.

*C.V. Theis, "The relation between the lowering of the piezometric surface and the rate and duration of discharge of a well using ground water storage," *Trans. Am. Geophys. Union, 16*, (1935), 519–524.

When $u < 0.01$, the Theis equation may be modified to*

$$s = h_0 - h = \frac{Q}{4\pi T}\left[-0.5772 - \ln \frac{r^2 S_c}{4Tt}\right]$$

$$= \frac{-2.30Q}{4\pi T} \log_{10} \frac{0.445r^2 S_c}{Tt} \tag{6.16}$$

The observed drawdown, $h_0 - h$, may be plotted against the pumping time t on a semi-logarithmic scale as shown in Figure 6.10.

Taking from the plotted data the value of $\Delta(h_0 - h)$ as the change in drawdown occurring over one log cycle of time, such as from 10 hr to 100 hr after pumping begins, the value of T can be calculated as follows:

$$\Delta(h_0 - h) = (h_0 - h_1) - (h_0 - h_2) = (h_2 - h_1)$$

$$= \frac{-2.30Q}{4\pi T}\left[\log_{10}\left(\frac{0.455r^2 S_c}{Tt_1}\right) - \log_{10}\left(\frac{0.445r^2 S_c}{Tt_2}\right)\right]$$

Figure 6.10. Jacob method for solution of the Theis equation (6.13).

*C.E. Jacob, "Drawdown test to determine effective radius of artesian well," Trans. ASCE, Vol. 112 (1947), 1047–1070.

or

$$T = \frac{2.30Q}{4\pi(h_1 - h_2)} \log_{10} \left(\frac{t_2}{t_1}\right) \tag{6.17}$$

In the semi-logarithmic scale, Eq. (6.16) is represented by a straight line which intersects the horizontal axis at a point designated as t_0. At this point, $(h_0 - h) = 0$ and we may write:

$$\log_{10} \frac{0.445r^2S_c}{Tt_0} = 0$$

or,

$$\frac{0.445r^2S_c}{Tt_0} = 1$$

Hence,

$$S_c = \frac{2.246Tt_0}{r^2} \tag{6.18}$$

Example 6.3

Using the Jacob's method described above, determine the transmissibility, T, the permeability K, and the storage constant S_c of the aquifer based on the data in Figure 6.10. The well pumps at a constant rate of 8.5 m³/hr, in an aquifer estimated to be 40 m thick, and the observation well is located at 20 m from the pumping well.

Solution

The drawdown observed during one cycle of time, $t_1 = 0.5$ hr and $t_2 = 5.0$ hr, is 0.9 m. Applying Eq. (6.17), the transmissibility T is

$$T = \frac{2.3 \cdot 8.5}{4\pi \cdot 0.9} \cdot \log_{10} (5.0/0.5) = 1.73 \text{ m}^3/\text{hr/m}$$

The permeability K is

$$K = T/b = 1.73/40 = 0.043 \text{ m/hr}.$$

The storage constant S_c can be estimated by using Eq. (6.18).

$$S_c = \frac{2.246 \cdot 1.73 \cdot 1.72 \cdot 10^{-2}}{(20)^2} = 1.67 \cdot 10^{-4}$$

6.4 AQUIFER BOUNDARIES

When a well is located close to an aquifer boundary, the shape of the drawdown curve may be significantly modified, which, in turn, affects the discharge rate as predicted by the radial flow equations.

The solution of aquifer boundary problems can often be simplified by applying the *method of images*. A hydraulic image is an imaginary source or sink with the same strength of the original well placed on the opposite side of the boundary to represent the effect of the boundary. Figure 6.11 shows the effect of an impermeable boundary on a well located at a close distance from it. Figure 6.12 shows the application of the method of images by placing an imaginary well of the same strength, Q, at the same distance on the opposite

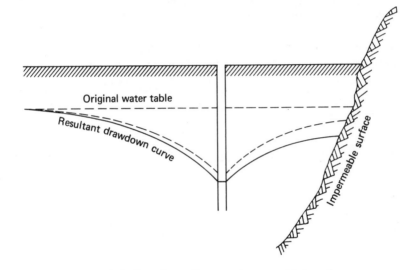

Figure 6.11. Pumping well near an impermeable boundary.

Figure 6.12. Equivalent hydraulic system with imaginary well.

side of the boundary. The original boundary conditions are thus replaced hydraulically by the two-well system in a hypothetically uniform aquifer of infinite extent. The two wells pumping at the same discharge rate hydraulically affect each other in such a way that the drawdown curves are linearly superimposed. The imaginary well and the real well offset one another at the boundary line, causing a zero hydraulic gradient ($dh/dr = 0$); hence, there is no flow across the boundary.

The presence of a lake, a river, or other similarly large body of water in the vicinity of a well, increases flow to the well. The effect of such a water body on the drawdown is exactly opposite to that of an impermeable boundary. Instead of an imaginary pumping well, the equivalent hydraulic system involves an imaginary *recharge well* placed at an equal distance across the boundary. The recharge well infuses water at a discharge rate, Q, into the aquifer under positive pressure. The resulting drawdown curve is obtained by linearly superimposing the drawdown component of the real well and the drawdown component of the imaginary recharge well that replaces the water boundary, as shown in Figures 6.13 and 6.14. The resulting drawdown curve

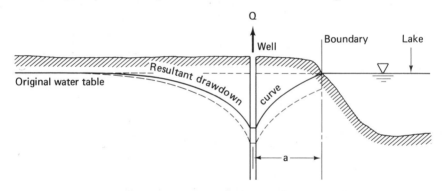

Figure 6.13. Pumping well near a perennial water body.

Figure 6.14. Equivalent hydraulic system with imaginary recharge well.

of the real well intersects the boundary line at the elevation of the free water surface. The steeper hydraulic gradient causes more water to flow across the boundary line.

Replacing the actual boundaries by an equivalent hydraulic system with imaginary wells can be applied to a variety of ground water boundary conditions. Figure 6.15(a) shows a discharge well pumping water from an aquifer bounded on two sides by impermeable boundaries. A system of three imaginary wells are required to provide the equivalent flow. The imaginary discharge wells, I_1 and I_2, provide the required absence of flow across the boundaries; a third imaginary well, I_3, is necessary to balance the system. The three imaginary wells all have the same discharge rate (Q) as the real well. All wells are placed at equal distances from the physical boundaries.

Figure 6.15(b) represents the situation of a discharge well pumping water from an aquifer bounded on one side by an impermeable boundary and on the other side by a perennial stream. The equivalent hydraulic system includes three imaginary wells of equal strength Q. I_1 is a recharge well and I_2 is a discharge well. A third imaginary well, I_3, a recharge well like I_1, is necessary to balance the system.

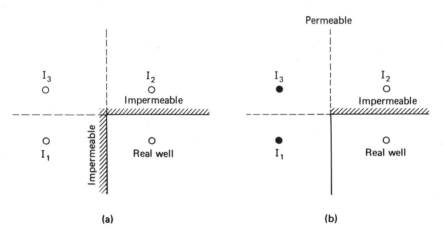

Figure 6.15. Discharging well near multiple boundaries.

Example 6.4

A factory located near a river bank wishes to extract a discharge of $0.04 \text{ m}^3/\text{s}$ from a confined aquifer ($K = 0.00012 \text{ m/s}$ and $b = 20 \text{ m}$). Local authorities require that at a distance of 30 m from the bank the ground water table may not be lower by more than 0.1 m from the surface elevation in the river. Determine the minimum distance from the bank that the well should be located.

Solution

As shown in Figures 6.13 and 6.16, a pumping well located near a perennial water body, such as a river, may be hydraulically replaced by an imaginary recharging well of the same strength on the opposite side and at the same distance from the boundary. The resultant drawdown curve may be obtained by superimposing the real and the imaginary wells and assuming an infinite extent of the aquifer without the boundary.

Assume P to be the distance between the pumping well and the river bank. Then the drawdown (at 30 m from the bank or $P - 30$ m from the pumping well and $P + 30$ m from the discharge well) is the sum of the piezometric surfaces at the point from the pumping well and the recharging well.

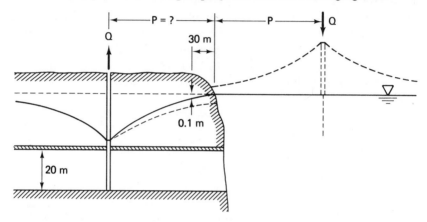

Figure 6.16

Rearranging Eq. (6.12), we have

$$h_2 - h_1 = s_1 - s_2 = \frac{Q}{2\pi bK} \ln\left(\frac{r_2}{r_1}\right)$$

$$(s_1 - s_2)_d = \frac{Q_{\text{disch}}}{2\pi bK} \ln\left(\frac{P}{P-30}\right) \qquad \text{(discharge well)}$$

$$(s_1 - s_2)_r = \frac{Q_{\text{rech}}}{2\pi bK} \ln\left(\frac{P+30}{P}\right) \qquad \text{(recharge well)}$$

Superimposing the two wells, as $s_{2_d} = s_{1_r}$, we have

$$s_{1_d} - s_{2_r} = \frac{Q}{2\pi bK}\left(\ln\frac{P}{P-30} + \ln\frac{P+30}{P}\right) = \frac{Q}{2\pi bK} \ln\left(\frac{P+30}{P-30}\right)$$

$$0.1 = \frac{0.04}{2\pi(20)(0.00012)} \ln\left(\frac{P+30}{P-30}\right)$$

and

$$\ln \left(\frac{P + 30}{P - 30}\right) = 0.0377$$

or

$$P + 30 = 1.038(P - 30)$$

Finally,

$$P = 1592 \text{ m}$$

PROBLEMS

6.4.1. A 30-cm diameter well penetrates the thickness of a 40-m-thick aquifer that has a coefficient of permeability of 0.6 m/h. The well is pumped at a constant rate of 0.10 m³/s. Determine the profile of the drawdown curve. If two such wells of the same strength are placed at the distance of 20 m apart, what would be the drawdown curve?

6.4.2. If the well in Prob. 6.4.1. were placed 20 m from a straight, impermeable aquifer boundary, determine the profile of the drawdown curve.

6.5   INVESTIGATION OF GROUND WATER

Locating subsurface ground water by information obtained on the earth's surface has been an ancient art known as *divination*. At the turn of this century geophysical methods had been developed for petroleum and mineral explorations. A few of these methods have proved useful for locating and analyzing ground water. Information obtained by surface methods can only provide indirect indication of ground water. Correct interpretation of the data usually requires supplemental information which can only be obtained by subsurface investigations. Two of the most commonly used geophysical methods are described below.

The Electrical Resistivity Method   Electrical resistivity of rock formations varies over a wide range. The measured resistivity of a particular formation depends on a variety of physical and chemical factors, such as the material and the structure of the formation, the size, shape, and distribution of pores, and the water content. The distinguishable difference between a rock formation alone and the same formation with large amounts of water filling the interstitial spaces is the key in detecting ground water.

The procedure involves measuring the electrical potential difference between two electrodes placed in the ground surface. When an electric current is applied through two other electrodes outside but along the same line with the potential electrodes, an electrical field penetrates the ground and forms a current flow network, as shown in Figure 6.17.

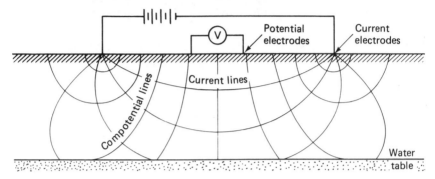

Figure 6.17. Schlumberger arrangement for resistivity determination.

A deeper penetration of the electrical field will occur by increasing the spacing between the electrodes. A variation in apparent resistivity is plotted against the electrode spacing, from which a smooth curve can be drawn.*

The interpretation of such a resistivity-spacing curve in terms of subsurface formation is frequently complex, and often difficult. Nevertheless, with certain supplemental data from subsurface investigations to substantiate the surface measurement, correct estimations can often be made.

The Seismic Refraction Method By shocking the earth surface with a small dynamite explosion or by the impact of a heavy weight, the time required for the sound or shock wave to reach a certain point at a known distance away can be measured. Seismic waves propagate through a medium the same way that light waves do. They may be refracted or reflected at the interface of any two materials of different elastic properties. A change in propagation velocity takes place at the interface. The wave speed is greatest in solid igneous rock and is the least in unconsolidated formations. The water content in a particular formation will significantly alter the wave speed in the formation. As the refraction wave is traveling several hundred meters deep into the ground, subsurface information may be obtained by placing several seismometers at various distances from the shock point along the same line. The wave travel time is plotted against the distance, as shown in Figure 6.18. The sudden change in the slopes of the time-spacing line can be interpreted to determine the depth of a ground water table.

*H. M. Mooney, and W. W. Wetzel, *The Potentials About a Point Electrode and Apparent Resistivity Curves for a Two-, Three-, and Four-Layered Earth* (Minneapolis: University of Minnesota Press, 1956).

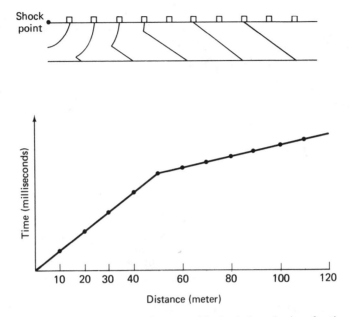

Figure 6.18. Determination of water table depth by seismic refraction method.

6.6 SEA WATER INTRUSION IN COASTAL AREAS

Along the coastline the coastal aquifers are in contact with sea water. Under natural conditions, fresh ground water is discharged into the sea under the water table, as shown in Figure 6.19. However, with an increased demand of ground water in certain coastal regions, the seaward flow of fresh ground water has been reduced or even reversed, causing saltwater from the sea to enter and penetrate the inland aquifers. This phenomenon is commonly known as *sea water intrusion*. If the saltwater travels far enough inland and enters the water supply wells, the ground water supply becomes useless. Furthermore, once the coastal aquifer is contaminated by salt, it is most difficult to remove the salt from the formation and the aquifer may be permanently damaged. Engineering prevention and control of sea water intrusion will be discussed in this section.

Overdrafting of coastal aquifers results in lowering the water table in unconfined aquifers or the piezometric surface in confined aquifers. The natural gradient originally sloping toward the sea is reduced or reversed. Because of the difference in densities of saltwater and freshwater, an interface

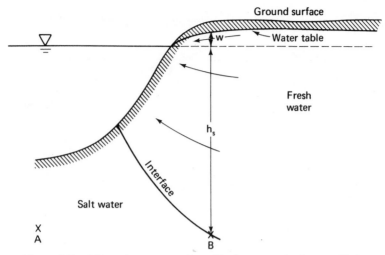

Figure 6.19. Schematic representation of freshwater and saltwater distribution in unconfined coastal aquifers.

is formed when the two liquids are in contact. The shapes and movements of the interface are governed by the pressure balance of the saltwater on one side and freshwater on the other side of the interface.

It was found that the interface occurring underground does not take place at sea level but at a depth below the sea level of approximately 40 times the height of the freshwater table above the sea level, as shown in Figure 6.19. This distribution is due to the equilibrium of hydrostatic pressure existing between the two liquids of different density.

Figure 6.19 shows the cross section of a coastal aquifer. The total hydrostatic pressure at point A at a depth h_s below the sea level is

$$P_A = \rho_s g h_s$$

where ρ_s is the density of the salt water and g is the gravitational acceleration. Similarly, the hydrostatic pressure at a point inland, B, at the same depth as A, and on the interface is

$$P_B = \rho_f g h + \rho_f g h_s$$

where ρ_f is the density of freshwater. For a stationary interface, the pressure at A and B must be the same, and we may write

$$\rho_s g h_s = \rho_f g h + \rho_f g h_s \tag{6.19}$$

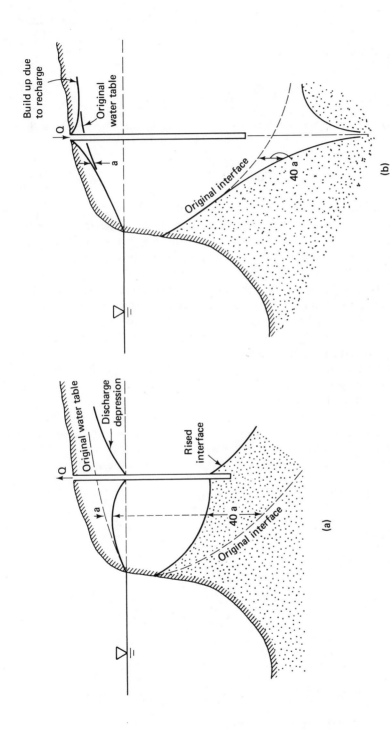

Figure 6.20. Sea water intrusion under the influence of (a) discharge and (b) recharge.

221

Solving Eq. (6.19) for h_s yields

$$h_s = \frac{\rho_f}{\rho_s - \rho_f} h \qquad (6.20)$$

By taking $\rho_s = 1.025$ g/cm² and $\rho_f = 1.000$ g/cm², the above relationship gives

$$h_s = \frac{1.000}{1.025 - 1.000} h = 40\,h \qquad (6.21)$$

This is commonly known as the *Ghyben–Herzberg relation*.

This relation shows that a small depression in the water table near the coastline due to well pumping could cause a big rise in the interface. Similarly, a buildup on the water table near the coastline due to artificial recharge could drive the saltwater wedge deep into the ground, thus forcing it to move seaward. These phenomena are schematically demonstrated in Figure 6.20.

Obviously, artificial recharge of an overdrafted coastal aquifer is an effective method of controlling sea water intrusion. By proper management, artificial recharge of the aquifer would eliminate the overdraft and maintain the proper water table level and gradient.

In addition to the artificial recharge, several other methods have also been effectively applied for control of sea water intrusion. The most common methods are described below.

1. *Pumping Trough:* A pumping trough is a line of discharge wells situated along the coastline. By pumping the wells, a depression trough line is formed, as shown in Figure 6.21. While saltwater is taken into the wells,

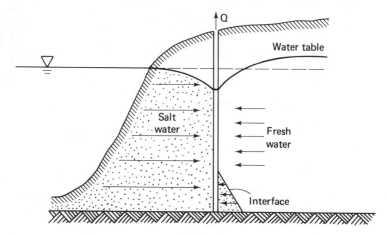

Figure 6.21. Control of sea water intrusion by pumping trough.

a certain amount of freshwater in the aquifer is also removed. The freshwater motion is in the seaward direction toward the wells. This movement of fresh ground water can stabilize the saltwater and freshwater interface.

2. *Pressure Ridge:* A pressure ridge is a series of recharge wells installed in parallel to the coastline. Freshwater is pumped into the coastal aquifer to maintain a freshwater pressure ridge along the coastline to control the saltwater intrusion. The pressure ridge should be sufficient to repel the sea water and should be located far enough inland; otherwise, the saltwater inland of the ridge will be driven farther inland, as demonstrated by Figure 6.22. Inevitably, a small amount of freshwater would

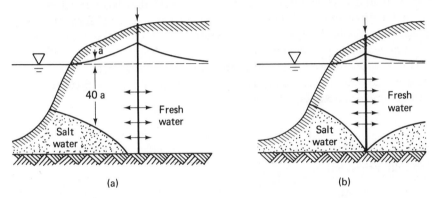

Figure 6.22. Control of sea water by pressure ridge. (a) Proper recharge arrangement. (b) Improper recharge arrangement.

be wasted to the sea; the remainder moving landward can be used to supply part of the pumping draft. Reclaimed waste water may be used to meet part of the recharge need. The method has the advantage of not depleting the usable ground water capacity, but the disadvantages of high initial and operating costs and the need of supplemental freshwater often make small-scale operations impractical.

3. *Subsurface Barriers:* Subsurface barriers may be built along the coastline to reduce the coastal aquifer permeability. In relatively shallow-layer aquifers, subsurface dikes may be constructed with sheet piling, puddled clay, or even concrete materials. An impermeable subsurface barrier may be formed by injecting flowable materials such as slurry, silicone gel, or cement grout into the aquifers through a line of holes. Subsurface barriers are best suited for locations such as a narrow, alluvial canyon connecting to a large inland aquifer. Although the initial cost of installing the barriers may be very high, there is almost no operation or maintenance expense.

6.7 SEEPAGE THROUGH DAM FOUNDATIONS

Dams constructed to store water in a reservoir may continuously lose water through seepage. Impermeable concrete dams constructed on an alluvial foundation may lose water through foundation seepage, while earth dams may also lose water through the dam body. Water movement in seepage flow is governed by Darcy's law in the same way as for ground water. Seepage flow can best be analyzed by applying the *flow net* technique.

A flow net is a graphical representation of flow patterns expressed by a family of *streamlines* and their corresponding *equipotential lines*. Streamlines are always drawn in the direction of the flow and they divide the flow field into a certain number of flow channels, each carrying the same discharge. Equipotential lines connect all points in the field that have equal velocity potential. The two sets of lines always meet at right angles and form an orthogonal net in the field. Figure 6.23 represents a portion of a flow net formed by streamlines and equipotential lines.

Flow nets are usually constructed with the distance between a pair of adjacent streamlines, Δn, and that between a pair of equipotential lines, Δs, in the same square in equal length.

$$\Delta n = \Delta s$$

The approximate velocity at any point in the field can be approximated by the following relation:

$$v = \frac{\Delta q}{\Delta n} \propto \frac{\Delta h}{\Delta s}$$

Darcy's law for steady flow at a low rate through a porous medium is

$$v = K\frac{dh}{ds} = K\frac{\Delta h}{\Delta s}$$

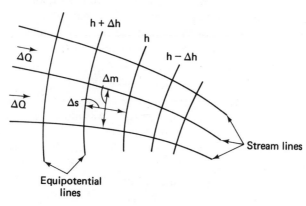

Figure 6.23. Flow net.

The rate of flow through a stream channel, per unit width of the dam, is

$$\Delta q = KA \frac{\Delta h}{\Delta s} \tag{6.22}$$

where A is the width of the square, or

$$A = \Delta n \tag{6.23}$$

Substituting Eq. (6.23) into Eq. (6.22), and $\Delta n = \Delta s$, we have

$$\Delta q = K(\Delta n) \frac{\Delta h}{\Delta s} = K \, \Delta h \tag{6.24}$$

Since Δh is a constant value of potential drop between any two adjacent equipotential lines, we may write

$$\Delta h = \frac{H}{n} \tag{6.25}$$

H is the difference between the reservoir water and tail water levels (Figure 6.24) and n is the number of squares in each stream channel of the flow net. The above equation may be rewritten as

$$\Delta q = K \frac{H}{n} \tag{6.26}$$

For m channels counted in a graphic flow net, the total flow rate per unit width of the dam is

$$q = m \, \Delta q = K \left(\frac{m}{n} \right) H \tag{6.27}$$

The total seepage discharge can be calculated by merely determining the ratio of m/n from a graphically constructed flow net.*

Figure 6.24 shows an example of seepage through a concrete dam without a cut-off wall and with a cut-off wall. A *cut-off wall* is a thin layer of impermeable material or sheet piling partially penetrating the aquifer under the dam. Depending on its location relative to the base of the dam, the cut-off wall alters the seepage flow pattern by lengthening the seepage paths, thus producing a vertical drop in the piezometric gradient. The cutoff walls effectively decrease the seepage rate and reduce the total uplift force on the base of the dam.

*Methods for construction of a flow net are presented in Appendix A.

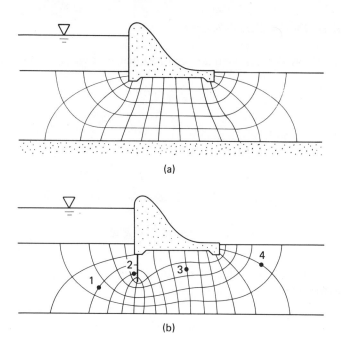

(a)

(b)

Figure 6.24. Seepage through pervious dam foundation. (a) Without cut-off wall. (b) With cut-off wall.

Example 6.5

The concrete gravity dam built on an alluvial channel bed as shown in Figure 6.24, stores water at a depth of 50 m. If the coefficient of permeability is $K = 2.14$ m/day, estimate the seepage per meter width of dam for (a) unmodified aquifer and (b) dam with a cut-off wall.

Solution

From Eq. (6.27)

$$Q = K \left(\frac{m}{n}\right) H$$

(a) For the unmodified aquifer without the cut-off wall, we count the number of flow channels, $m = 5$, and the number of equipotential drops, $n = 11$, in Figure 6.24(a).

$$Q = 2.14 \cdot \frac{5}{11} \cdot 50 = 48.64 \ m^3/\text{day/m width}$$

(b) For modified aquifer with cut-off wall, as shown in Figure 6.24(b), we count the number of flow channels, $m = 5$, and the number of equipotential drops, $m = 16$. Therefore,

$$Q = 2.14 \cdot \frac{5}{16} \cdot 50 = 33.44 \ m^3/\text{day/m width}$$

PROBLEMS

6.7.1. If the water elevation in the upstream reservoir of Example 6.4 is lowered to 40 m, determine the seepage rate per unit width of the dam foundation. Estimate the magnitude and direction of the velocity at locations 1, 2, 3, and 4.

6.7.2. Determine the required length of an impervious floor at the upstream heel of the dam, x, in Figure 6.25 to reduce the quantity of foundation seepage by 50%.

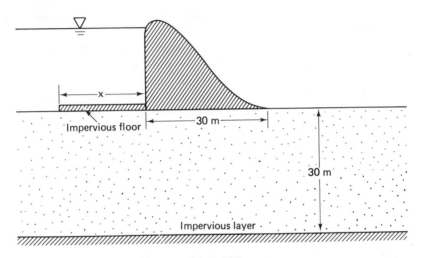

Figure 6.25

6.7.3. Determine the quantity of seepage per unit width of the channel bed in Figure 6.26. The permeability $K = 0.00051$ cm/s.

Figure 6.26

227

6.8 SEEPAGE THROUGH AN EARTH DAM

Since an earth dam is built with pervious materials, it is of particular engineering concern, for it may be damaged by seepage through the dam body, for example, by piping. Seepage analysis should be made for each earth dam designed by applying the method of flow nets.

Seepage through an earth dam body can be treated as flow through an unconfined porous medium. The upper surface of the flow, known as the *surface of saturation* or *phreatic surface*, is at atmospheric pressure. The typical shape of a *phreatic line* in a homogeneous earth dam is shown in Figure 6.27. The phreatic line is a streamline whose intersection with the equipotential lines is equally spaced vertically by the amount of $\Delta H = H/n$, where H is the total head available and n is the number of potential drops counted in a graphic flow net. This line, which provides the upper boundary of the flow net, must be initially located by trial. An empirical rule for locating the phreatic line was suggested by Casagrande* and is shown in Figure 6.27.

Section DF on the lower part of the downstream dam face must be protected against soil piping which may eventually lead to dam failure. The seepage water may be removed permanently from the downstream surface by a properly designed drainage system. For a nonstratified, homogeneous earth dam, a narrow, longitudinal drain will effectively intercept all the water seeping through the embankment. Figure 6.28 shows schematically the dimensions of a typical earth dam drainage.

The total discharge through an earth dam can be determined by the graphically constructed flow net, as discussed in Section 6.7. Equation (6.27) gives the seepage discharge for the unit width of the dam.

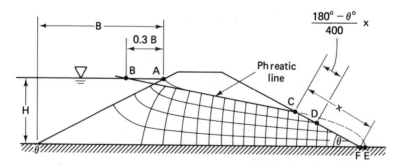

Figure 6.27. Seepage flow net through a homogeneous earth dam. (A large portion of the phreatic line AD can be approximated by the parabola BCE with F as focus and passing through the point B. Point A on the upstream face of the dam is the intersection of the water surface with the dam. Point D is the downstream transition where the seepage is exposed to the atmosphere.)

*A. Casagrande, "Seepage Through Dams," *J. New Eng. Water Works Assoc., 51* (1937), 139.

$$Q = K\left(\frac{m}{n}\right)H$$

where m is the number of flow channels and n is the number of equipotential drops in the flow net.

Figure 6.28. Drainage blanket in an earth dam.

PROBLEMS

6.8.1. An earth dam (see Figure 6.29) is 30 m high and has a 3-m freeboard and a 3-m crest width. The upstream face of the dam has a 1(V): 2(H) slope and a 1(V): 3(H) downstream face slope. The dam material has a permeability coefficient of 0.0001 cm/s. Compute the seepage per unit width from the network based on the formula of the equivalent parabola.

Figure 6.29

6.8.2. An earth dam (see Figure 6.30) is constructed with a uniform material having a coefficient of permeability $K = 2 \cdot 10^{-4}$ m/s on a relatively impervious foundation. The dam is 30 m high and has a top width of 15 m. The upstream face of the dam has a 1(V): 1(H) slope and a downstream face of 1(V): 2(H). Compute the seepage rate in units of cubic meters per day in unit width of the dam.

Figure 6.30

6.8.3. Determine the seepage rate per unit width of the dam in Prob. 6.8.2 if a drainage blanket extends back 10 m from the toe of the dam.

7

HYDRAULIC STRUCTURES

7.1 FUNCTIONS OF HYDRAULIC STRUCTURES

Water is useful to humans only if it is properly controlled, conveyed, and contained. Hydraulic structures are structures made by humans and designed to serve these purposes. Because of the great variety of hydraulic structures that can be built to serve any one or more purposes, any classification of hydraulic structures will inevitably be discretionary, and there are no clearly defined general classifications. For example, a general classification of hydraulic structures based on use is not satisfactory because many identical structures may be used to serve completely different purposes. For example, a low dam could be built across a channel as a device to measure discharge or a low dam could be built to raise the water level at the entrance to an irrigation canal to permit diversion of water into the canal. Instead of classifying the various hydraulic structures into arbitrary catergories, we have listed below the common functions of hydraulic structures and the basic design criteria:

1. Storage structures are designed to store water under hydrostatic conditions. A storage structure usually has a large capacity for a relatively small change in hydrostatic head (water elevation).
2. Conveyance structures are designed to convey water from one place to

another. The design normally emphasizes delivery of a given discharge with a minimum consumption of energy.

3. Waterway and navigation structures are designed to ensure water transportation. The protection and maintenance of a minimum water depth throughout the entire waterway under various possible conditions are emphasized in the design.

4. Measurement or control structures are used to determine the discharge in a particular conduit. Stable performance and a one-to-one relationship between the indicator and the corresponding discharge are necessary.

5. Energy conversion structures are designed to convert hydraulic energy into mechanical or electric energy (e.g., hydraulic turbine systems) or to convert electrical or mechanical energy into hydraulic energy (e.g., hydraulic pumps). The design emphasis is on the efficiency of the system.

6. Sedimentation or fish control structures are designed to direct or control the movement of the nonhydraulic elements in water. An understanding of the basic mechanisms and behaviors of the elements involved is an essential requirement for the design.

7. Energy dissipation structures are used to dissipate the excess amount of water energy for prevention of erosion in a channel.

8. Collection structures are designed to collect and admit water to a conduit or system. A typical example is a surface drainage inlet used to collect storm runoff to a sewer system.

Obviously, detailed consideration of all these functions and their design criteria are beyond the scope of this book. Only the most commonly encountered hydraulic structures are discussed here to demonstrate the fundamental considerations in hydraulic structures.

7.2 STABILITY OF GRAVITY DAMS AND ARCH DAMS

A dam is a barrier structure placed across a watercourse to store or slow the normal water flow. Water stored behind a dam may reach a great height depending on the height of the dam. Low dams are usually a few meters in height; high dams are those over 100 meters in height. A typical gravity dam is a massive structure (Figures 7.1 and 7.2). The enormous weight of the dam body provides the necessary stability against tip-over about the toe of the dam or shear failure along a horizontal plane. Arch dams are normally built on solid rock foundations that provide resistance to the load by arch actions

Figure 7.1. Top view representation of a gravity dam.

Figure 7.2. Cross-sectional representation of a gravity dam.

(Figure 7.3). Generally speaking, low dams are usually gravity dams and high dams (100 m and higher) can be arch dams. Combining gravity action and arch action in dam design is a common practice. A dam that has a com-bination design may be called an *arch-gravity dam* or a *gravity-arch dam*, depending on the proportion of the actions involved.

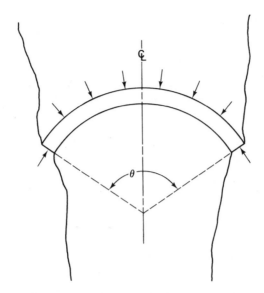

Figure 7.3. Top view representation of an arch dam and reactions.

The major forces acting on a gravity dam (or any other type of dam) are represented in Figure 7.2. They are:

1. Hydrostatic force, F_{HS}
2. Weight of the dam, W
3. Uplifting force on the base of the dam, F_u
4. Sedimentation or silt deposit pressure force, F_s
5. Earthquake force on the dam, F_{EQ}
6. Earthquake force due to the water mass behind the dam, F_{EW}

The straight-line structure of the gravity dam (Figure 7.1) allows force analysis per unit width of the dam. The binding forces between each unit width segment are neglected in the analysis because they can only add to the stability of the dam.

The hydrostatic force acting on the upstream face of the dam may be resolved into a horizontal component and a vertical component. The horizontal component of the hydrostatic force acts along a horizontal line $H/3$ above the base of the dam. This horizontal force forms a clockwise moment about the toe of the dam, and it is a major factor in the possible overturning of the dam. It may also cause dam failure by shear along a horizontal plane. The vertical component of the hydrostatic force is equal to the weight of the water mass directly above the upstream face of the dam. It acts along a vertical line that passes through the centroid of that mass. The vertical component of hydrostatic force always forms a counterclockwise moment about the toe. It is a major stabilizing factor in gravity dams.

The largest stability force is the weight of the dam which depends not only on the dimensions but also on the material used. The unit weight of most masonary or solid earth materials is approximately from 2.4 to 2.6 times that of water. The importance of this single stability force of gravity explains the name gravity dam.

The uplifting force on the base of a dam can be determined by foundation seepage analysis (Chapter 6). Acting in the opposite direction of the gravity, this force should be minimized in every dam design, if possible. It weakens the foundation by the uplifting and it tends to overturn the dam. If the foundation soil is porous and homogeneous in structure, the uplifting pressure on the base varies linearly from full hydrostatic head H at the heel of the dam to the full hydrostatic head of the tail water at the toe. The total resulting uplifting force can be determined by integrating the trapezoidal area represented by the dashed line in Figure 7.3. The magnitude of the uplifting force as well as the overturning (clockwise) momentum can be greatly reduced by installing an impermeable cut-off wall a few meters into the foundation. The cut-off wall alters the seepage course by lengthening the pathway, thus reducing the seepage and uplifting force.

The water flow velocity immediately behind the dam is very slow or near zero. The water loses its ability to carry sediments or other suspended matters. These heavier materials are deposited on the bottom of the lake near the base of the dam. The silt–water mixture is approximately 50% heavier than water (sp. gr. 1.5) and forms an excess pressure force near the heel. Normally, the thickness of the silt layer may increase slowly with time. This force may contribute to dam failure by shear along the base.

In earthquake zones the forces generated by earthquake motion must be incorporated into dam design. Earthquake force on dams is due to acceleration associated with the earthquake motions. The magnitude of the earthquake force in the dam body, F_{EQ}, is proportional to the acceleration and the mass of the dam body. The force may be acting in any direction through the centroid of the dam body.

Earthquake force due to acceleration of the water body behind the dam is approximately equal to

$$\frac{5}{9}\left(\frac{a\gamma}{g}\right)H^2$$

where a is the earthquake acceleration, γ is the specific weight of water, and H is the hydrostatic head, or depth of water, immediately behind the dam.

The earthquake force of the water body acts in a horizontal direction at distance $(4\pi/3)H$ above the base of the dam.*

In order to ensure stability, the safety factors of a dam against sliding failure or overturning must be greater than unity. In addition, the maximum pressure exerted on the foundation must not exceed the bearing strength of the foundation.

The *ratio of forces against sliding* is defined by the ratio of the total horizontal resistance force that the foundation can develop to the sum of all forces acting on the dam that tend to cause sliding. This ratio may be expressed as

$$\text{force ratio against sliding} = \frac{\mu(\Sigma F_v) + A_s \tau_s}{\Sigma F_H}$$

where μ is the coefficient of friction between the dam base and the foundation (ordinarily $0.4 < \mu < 0.75$), ΣF_v is the summation of all vertical force components acting on the dam, τ_s is the shear stress strength of keys, and A_s is the total shear area provided by the keys. ΣF_H is the summation of all horizontal force components acting on the dam.

The keys are the part of the dam body built into the foundation to add resistance against dam sliding. Horizontal forces are transmitted to the foundation via the shear force in the keys. The total shear force provided by the keys, $\tau_s A_s$, must be larger than the difference between the total horizontal force acting on the dam, ΣF_H, and the friction force provided by the base, $\mu(\Sigma F_v)$.

$$\tau_s A_s > [\Sigma F_H - \mu(\Sigma F_v)]$$

The *ratio of forces against overturning* is defined by the ratio of the resisting moments (counterclockwise moments about the toe) to the overturning moments (clockwise moments about the toe):

$$\text{force ratio against overturning} = \frac{W \cdot l_w + (F_{HS})_v \cdot l_v}{\Sigma F_H \cdot Y_H + F_u \cdot l_u}$$

where $(F_{HS})_v$ is the vertical component of the hydrostatic force and l_w, l_v, and l_u are the horizontal distances from the toe to the lines of action of the weight, the vertical component of hydrostatic force, and the uplifting force, respectively. Y_H is the vertical distance measured from the toe to the lines of action of each respective horizontal force component, F_H.

*J. I. Bustamante, "Water Pressure on Dams Subjected to Earthquakes," *J. Engr. Mech., Div.*, A.S.C.E., *92* (October, 1966), 115–127.

Figure 7.4. Pressure distribution on dam foundation.

The vertical pressure on the foundation may be assumed to be a linear distribution between the toe and heel as shown in Figure 7.4. If we let R_v represent the resultant of all vertical forces acting on the base of the dam and P_t and P_H represent the pressure at the toe and heel, respectively, we may write

$$R_v = \frac{(P_T + P_H)}{2} \cdot B$$

and

$$R_v = \frac{(P_T - P_H)}{2}\left(\frac{B}{e}\right) \cdot \left(\frac{B}{6}\right)$$

Solving the above equations simultaneously, we have

$$P_T = \left(\frac{R_v}{B}\right)\left(1 + \frac{6e}{B}\right) \tag{7.1}$$

$$P_H = \left(\frac{R_v}{B}\right)\left(1 - \frac{6e}{B}\right) \tag{7.2}$$

The vertical resultant force normally acts through a point on the downstream side of the base centerline. Therefore, P_T is usually the critical pressure in design. The value of P_T must be kept less than the bearing strength of the foundation. The pressure at the heel, P_H, is of less importance. Nevertheless, it is desirable to keep P_H a positive value at all times in order to prevent tension cracks from developing in the heel region. Negative pressure indicates tension, and masonry materials have very low resistance to tension stress. A positive P_H value can be ensured if the vertical resultant force, R_v, is kept within the middle third of the base, or

$$e < 6B \tag{7.3}$$

The force load on an arch dam is essentially the same as on a gravity dam. To resist these forces, the dam foundation must provide a horizontal arch reaction. The large horizontal reactions can only be provided by strong, solid rock abutments at the two ends of the arch (Figure 7.3). Arch dams are mostly high dam construction and are built restrictively in relatively narrow rock canyon sections. The efficiency of using material strength, and the slim dam base in relation to the height of the dam, make the arch dams the unique choice for many situations. Since the arch dam combines the resistance from the arch action and the gravity, the high stress in each segment of the dam requires detailed stress analysis.

The stability analysis on an arch dam is usually carried out on each horizontal rib. Take a rib situated h m below the designed reservoir water level. The forces acting in the direction of the dam centerline may be summed up as follows:

$$2R \sin \frac{\theta}{2} = 2r\gamma h \sin \frac{\theta}{2}$$

where R is the reaction from the abutment, θ is the central angle of the rib, r is the outer radius (extrados) of the arch, and γh is the hydrostatic pressure acting on the rib. The above relation can be simplified for the abutment reaction:

$$R = r\gamma h \tag{7.4}$$

This value is determined by considering only the arch reaction for resisting the hydrostatic load on the dam. In practice, however, several other resistive forces, as discussed in the gravity dams, should also be included. The analysis should consider the combination resistance from both the arch and gravity actions.

The volume of an arch dam is directly related to the thickness of each rib, t, the width of the rib, B, and the center angle, θ. For the minimum volume of the dam, it can be shown that $\theta = 133° \ 34'$. Other factors, such as topographic conditions, often prevent the use of this optimal value. Values within the range of $110° < \theta < 140°$ are commonly used in arch dam design.

A simple approach to arch dam design is to keep constant the central angle while the radii vary from rib to rib, as shown in Figure 7.5. Another approach frequently used is to keep the radii of the ribs at a constant value and allow variations of the central angle, as shown in Figure 7.6.

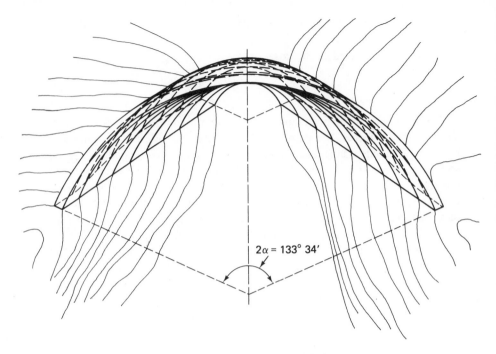

Figure 7.5. Constant-angle arch dam.

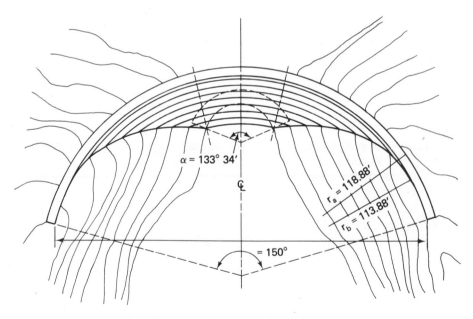

Figure 7.6. Constant-radius arch dam.

7.3 WEIRS

Any obstruction of a streamflow over which water flows can be called a *weir*. Since weirs are usually installed in stream flows that have free surfaces, the flow behavior over a weir is governed by gravity forces. As a result of a sudden reduction of area, water over the weir is accelerated to pass the crest. This fact has been utilized in the past to keep flood currents from overflowing a bridge, as schematically shown in Figure 7.7. By placing an obstruction of

Figure 7.7. Acceleration of flow over a weir.

adequate dimensions in an otherwise subcritical stream flow, the water level is raised upstream from the weir. With the increased available head, the flow is accelerated to pass over the crest. This acceleration causes the depth of water to reduce and reach the depth of supercritical flow after passing the weir. Some distance downstream from the weir the flow returns to a normal subcritical depth through a hydraulic jump. This arrangement makes the bridge structure free from the high flood water.

Flow acceleration over the weir provides an unique one-to-one relationship between the approaching water height and the discharge for each type of weir. Weirs are commonly built to measure the discharge in open channels. Weirs are also used to raise stream-flow levels in order to divert a desired amount of discharge.

The use of a weir as a flow-measuring device will be discussed in detail in Chapter 9. In this section the hydraulic characteristics of weirs will be presented.

A weir structure increases the depth immediately upstream of the weir and reduces the water cross section at the crest. The increase of water depth slows down the water velocity upstream; and the sudden reduction of cross-sectional area causes the flow to speed up hastily as it passes over the crest. The occurrence of critical flow on weirs is the essential feature of weir structures.

The hydraulics of an overflow weir may be examined by using an ideal frictionless weir (see Figure 7.8). At the occurrence of the critical depth the discharge per unit width of the weir can be determined by using the critical flow conditions:

$$\frac{V_c}{gd_c} = 1 \tag{5.11}$$

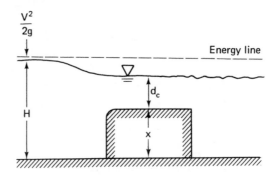

Figure 7.8. Flow over a frictionless weir.

and

$$d_c = \sqrt[3]{\frac{q^2}{g}} \tag{5.13}$$

Rearranging Eq. (5.11), we may write

$$\frac{V_c^2}{2g} = \frac{d_c}{2}$$

The energy at the critical section is

$$E = d_c + \frac{V_c^2}{2g} = d_c + \frac{d_c}{2} = \frac{3}{2}d_c \tag{5.8}$$

The energy of the approaching flow is approximately equal to the water depth upstream of the weir, H, since the velocity head is small. The same amount of energy is available at the critical section:

$$E + x = H = \frac{3}{2}d_c + x \tag{7.5}$$

Combining Eqs. (7.5) and (5.13) and defining $H_s = H - x$, we have

$$q = \sqrt{gd_c^3} = \sqrt{g\left(\frac{2H_s}{3}\right)^3} = 1.705H_s^{3/2} \tag{7.6}$$

This is the basic form of the weir equation. The coefficient 1.705 is higher than the coefficients obtained in experiments because friction loss is neglected in the above analysis.

A *sharp-crest weir* is shown in Figure 7.9. Upstream from the weir the velocity of all water elements is nearly uniform and parallel. The water elements near the bottom of the channel rise from the bottom in the immediate

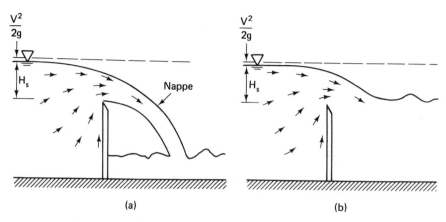

Figure 7.9. Water flows over sharp-crest weir. (a) Free-falling water nappe.
(b) Submerged flow.

neighborhood of the weir in order to pass the crest. The vertical component
of the flow near the upstream face of the weir causes the lower surface of the
stream to separate from the weir and form a *nappe* [see Figure 7.9(a)]. The
nappe usually traps a certain amount of air between the lower nappe surface
and the downstream side of the weir. If no means of air restoring are provided, a void will appear which represents a negative pressure on the structure. The nappe will also cling intermittently to the side of the weir and
cause the flow to be unstable. The dynamic effect of this unstable flow may
result in added negative pressure. This negative pressure may eventually
damage the structure.

When the downstream water level rises over the weir crest, the weir is said
to be *submerged* [see Figure 7.9(b)]. In this case, the negative pressure no
longer exists, and a new set of flow parameters may be considered in the
determination of the discharge coefficient (Chapter 9).

Example 7.1

Uniform flow occurs in a long rectangular channel 4 m wide at a depth of
2 m. The channel is laid on a slope of 0.001 and has a Manning's coefficient
$n = 0.025$. Determine the minimum height of a low weir that can be built on
the floor of this channel (see Figure 7.10) to produce critical depth.

Solution

For a uniform flow condition, the Manning's uniform flow formula [Eq.
(5.5)] may be used to determine the channel discharge, Q:

$$Q = \frac{1}{n} A R_H^{2/3} S^{1/2}$$

Figure 7.10

Here, $A = 2 \text{ m} \cdot 4 \text{ m} = 8 \text{ m}^2$, $P = 2 \cdot 2 \text{ m} + 4 \text{ m} = 8 \text{ m}$, and $R_H = A/P$
$= 1.0 \text{ m}$. Hence,

$$Q = \frac{1}{0.025}(8)(1.0)^{2/3}(0.001)^{1/2} = 10.12 \text{ m}^3/\text{s}$$

and

$$V = \frac{Q}{A} = \frac{10.12}{8} = 1.26 \text{ m/s}$$

The specific energy is

$$E = d + \frac{V^2}{2g} = 2 + \frac{(1.26)^2}{2 \cdot 9.81} = 2 + 0.082 = 2.082 \text{ m}$$

For a critical flow condition to occur, the condition of Eq. (7.5) must be
satisfied, or

$$E_c = \frac{3}{2}d_c = \frac{3}{2}\sqrt[3]{\frac{q^2}{g}}$$

Hence, $d_c = 0.87 \text{ m}$.
The corresponding velocity is

$$V_c = \frac{Q}{4d_c} = \frac{10.12}{4 \cdot 0.87} = 2.92 \text{ m/s}$$

and the critical velocity head is

$$\frac{V_c^2}{2g} = 0.43 \text{ m}$$

If there is no energy loss at the weir, the minimum height of the weir that
can be built in the floor to produce critical flow in this channel is x.

$$x = E - \left(d_c + \frac{V_c^2}{2g}\right) = 2.082 - (0.87 + 0.43) = 0.781 \text{ m}$$

PROBLEMS

7.3.1. If a rectangular channel carries a discharge $2 \text{ m}^3/\text{s}$ per unit channel width, and the energy line measured 2.7 m above the channel bottom and the weir crest is 1.4 m in height, what is the flow velocity over the weir? Determine the coefficient of discharge considering energy losses (in reference to Eq. 7.6).

7.3.2. The elevation of the crest of an overflow weir is 100 m above the mean sea level at a certain port. If the weir is 5 m long and discharges $10 \text{ m}^3/\text{s}$, determine the water surface elevation upstream from the weir. Neglect friction loss.

7.3.3. A weir 1.0 m in height is built across the floor of a 4-m-wide rectangular channel. If the water depth on the crest measures 0.3 m, determine the discharge and the water depth upstream from the weir.

7.3.4. The depth of water upstream from a 10-m-wide frictionless weir 1 m in height is 1.5 m. Determine the discharge.

7.4 OVERFLOW SPILLWAY

A spillway is an essential part of a large dam and provides an efficient, safe means of releasing flood water that exceeds the design capacity of the reservoir. Basically, an overflow spillway is an open channel on large slopes that allows the excess water to flow out at supercritical velocities. The ideal longitudinal profile of an overflow spillway should flow along the same curve as the underside of the free-falling water nappe (Figure 7.11) to minimize the pressure on the spillway surface. However, caution must be exercised to avoid any negative pressure on the surface. Negative pressure is caused by separation of the high-speed water flows over the spillway, resulting in a pounding action of the spillway surface that can cause significant damage to the spillway structure.

The U.S. Waterways Experimental Station suggests a set of simple crest profiles that have been found to be in good agreement with actual prototype measurements. The geometry of the U.S. Waterways Experimental Station spillway crest profiles is shown in Figure 7.12.

The discharge of a spillway may be calculated by a formula similar to that derived for flow over a weir, [Eq. (7.6)]:

$$Q = CLH_a^{3/2} \qquad (7.7)$$

where C is the coefficient of discharge, L is the width of the spillway crest, and H_a is the sum of the static head, H_s, and the approaching velocity head, $V_a^2/2g$, at the crest,

$$H_a = H_s + \frac{V_a^2}{2g} \qquad (7.8)$$

Figure 7.11. Ideal longitudinal profile of an overflow spillway. (a) Water nappe over sharp-crest weir. (b) Flow profile in overflow spillway.

		Upstream slope (vert/hor.)		
	3/0	3/1	3/2	3/3
a/H_s	0.175	0.139	0.115	0
b/H_s	0.282	0.237	0.214	0.199
r_1/H_s	0.50	0.68	0.48	0.45
r_2/H_s	0.20	0.21	0.22	–
K	0.500	0.516	0.515	0.534
P	1.850	1.836	1.810	1.776

$$\left(\frac{y}{H_s}\right) = -K\left(\frac{x}{H_s}\right)^P$$

Figure 7.12. Overflow spillway profile.

The coefficient of discharge of a particular spillway crest is often determined by scaled model tests (Chapter 8). The value of the coefficient normally ranges between 1.66 to 2.26. A detailed discussion of Eq. (7.7) will be presented in Chapter 9.

Example 7.2

An overflow spillway 80 m wide carries a maximum discharge of 400 m³/s. Neglecting the approaching velocity head, define the crest profile for the spillway. Consider a 3:1 upstream slope and a 2:1 downstream slope.

Solution

Applying Eq. (7.7), we get

$$Q = 2.22LH_a^{3/2} \cong 2.22LH_s^{3/2}$$

and

$$H_s = \left(\frac{Q}{2.22L}\right)^{2/3} = \left(\frac{400}{2.22 \cdot 80}\right)^{2/3} = 1.72 \text{ m}$$

From the table in Figure 7.12 we have

$$a = 0.139H_s = 0.239 \text{ m}; \qquad r_1 = 0.68H_s = 1.170 \text{ m}$$
$$b = 0.237H_s = 0.408 \text{ m}; \qquad r_2 = 0.21H_s = 0.361 \text{ m}$$
$$K = 0.516; \qquad P = 1.836$$

and

$$\left(\frac{V}{H_s}\right) = -K\left(\frac{x}{H_s}\right)^P = -0.516\left(\frac{x}{H_s}\right)^{1.836}$$

The downstream end of the profile curve will be matched to a straight line with slope 2:1. The position of the point of tangency is determined by

$$\frac{d\left(\frac{y}{H_s}\right)}{d\left(\frac{x}{H_s}\right)} = -KP\left(\frac{x}{H_s}\right)^{P-1} = -0.947\left(\frac{x}{H_s}\right)^{0.836} = -2$$

Hence,

$$\frac{X}{H_s} = 2.45, \qquad X_{P.T.} = 4.21 \text{ m}$$

$$\frac{Y}{H_s} = 2.674, \qquad Y_{P.T.} = 4.599 \text{ m}$$

The crest profile curve of the spillway is shown in Figure 7.13.

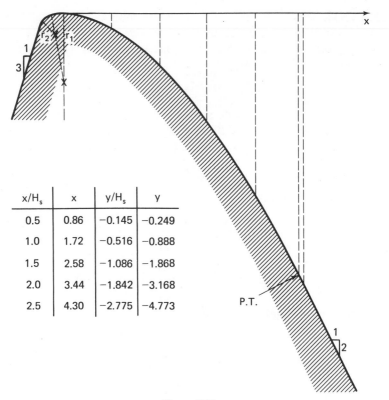

x/H_s	x	y/H_s	y
0.5	0.86	−0.145	−0.249
1.0	1.72	−0.516	−0.888
1.5	2.58	−1.086	−1.868
2.0	3.44	−1.842	−3.168
2.5	4.30	−2.775	−4.773

Figure 7.13

PROBLEMS

7.4.1. An overflow spillway is designed to discharge 230 m³/s and has a maximum head of 2.2 m. Determine the width and the profile of the spillway crest having a vertical upstream slope and a 1.5 : 1 downstream slope.

7.4.2. Determine the maximum discharge for an overflow spillway 38 m wide and having an available head of 1.86 m. Define the crest profile for both the upstream and downstream slopes of 1 : 1.

7.5 SIDE-CHANNEL SPILLWAYS

A side-channel spillway carries water from an overflow crest in a channel parallel to the crest (Figure 7.14).

The discharge per unit width of the overflow crest can be determined by Eq. (7.6), and the discharge through any section of the side-channel at dis-

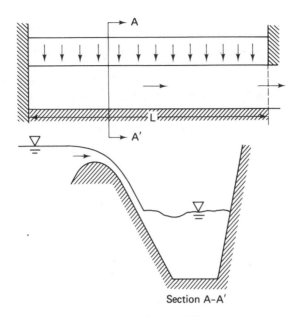

Figure 7.14.   Side-channel spillway.

tance x from the upstream end of the channel is

$$Q_x = xCH^{3/2} \tag{7.9}$$

The side-channel spillway must provide a slope steep enough to carry away the accumulating flow in the channel, but a minimum slope and depth at each point along the channel is desired in order to minimize construction costs. For this reason, an accurate water surface profile for the maximum design discharge is important in the side-channel spillway design.

The flow profile in the side-channel cannot be analyzed by the energy principle due to the highly turbulent flow conditions that cause excess energy loss in the channel. However, an analysis based on the momentum principle has been validated by both model and prototype measurements.

The momentum principle considers the forces and change of momentum between two adjacent sections, dx distance apart, in the side-channel:

$$\Sigma F = \rho(Q + \Delta Q)(V + \Delta V) - \rho QV \tag{7.10}$$

where ρ is the density of water, V is the longitudinal velocity, and Q is the discharge at the upstream section. The symbol Δ signifies the incremental quantity at the adjacent downstream section.

The forces represented on the left-hand side of Eq. (7.10) usually include the weight component of the water body between the two sections in the direction of the flow, $(\rho g A\, dx) \sin \theta$; the unbalanced hydrostatic forces,

$$\rho g A \bar{d} \cos \theta - \rho g (A + \Delta A)(\bar{d} + \Delta \bar{d}) \cos \theta$$

and a friction force, F_f, on the channel bottom. Here A is the water cross-sectional area, $\bar{d}$ is the distance between the centroid of the area to the water surface, and θ is the angle of the channel slope.

The momentum equation may thus be written as

$$\rho g A \Delta x \sin \theta + [\rho g A \bar{d} - \rho g (A + \Delta A)(\bar{d} + \Delta \bar{d})] \cos \theta - F_f$$
$$= \rho (Q + \Delta Q)(V + \Delta V) - \rho Q V \qquad (7.11)$$

Let $S_0 = \sin \theta$ for a reasonably small angle, and $Q = Q_1$, $V + \Delta V = V_2$, $A = (Q_1 + Q_2)/(V_1 + V_2)$, and $F_f = \gamma A S_f\, \Delta x$; the above equation may be simplified to

$$\Delta d = -\frac{Q_1(V_1 + V_2)}{g(Q_1 + Q_2)}\left(\Delta V + V_2 \frac{\Delta Q}{Q_1}\right) + S_0\, \Delta x - S_f\, \Delta x \qquad (7.12)$$

where Δd is the change in water surface elevation between the two sections. This equation is used to compute the water surface profile in the side-channel. The first term on the right-hand side represents the change in water surface elevation between the two sections due to the impact loss caused by the water overfalling into the channel. The second term represents the change due to friction in the channel. Relating the water surface profile to a horizontal datum, we may write

$$\Delta z = \Delta d - S_0\, \Delta x = -\frac{Q_1(V_1 + V_2)}{g(Q_1 + Q_2)}\left(\Delta V + V_2 \frac{\Delta Q}{Q_1}\right) - S_f\, \Delta x \qquad (7.13)$$

It may be noticed that when $Q_1 = Q_2$ or when $\Delta Q = 0$, Eq. (7.13) reduces to

$$\Delta z = \left(\frac{V_2^2}{2g} - \frac{V_1^2}{2g}\right) - S_f\, \Delta x \qquad (7.14)$$

which is the energy equation for constant discharge in an open channel as derived in Chapter 5.

PROBLEMS

7.5.1. The overflow spillway of Example 7.2 discharges into a side-channel spillway 5 m wide with horizontal bottom slope. If the wall opposite to the overflow spillway crest is vertical and the depth of water at the exit end of the side-channel is the critical depth, determine the depth of water at the beginning (the upstream end) of the channel. The width of the channel bottom is 10 m.

7.5.2. Solve Prob. 7.5.1 if the channel bottom slope is 0.004.

7.5.3. A rectangular side-channel spillway 100 m long discharges 5.0 m³/s/m. The channel is 6 m wide and has a bottom slope of 0.0009. Define the water surface profile for a critical depth at the exit end of the channel as shown schematically in Figure 7.15.

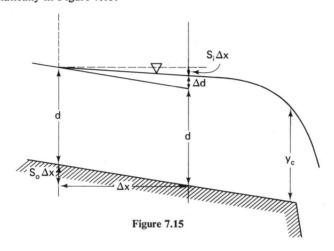

Figure 7.15

7.6 SIPHON SPILLWAYS

When a closed conduit is elevated above the hydraulic gradient line (pressure line), the conduit experiences negative pressure. A spillway designed to discharge water in a closed conduit under negative pressure is known as a *siphon spillway*. A siphon spillway begins to discharge under negative pressure when reservoir water level reaches a certain elevation which primes the conduit. At the beginning stage the water overflows the spillway crest in the same manner as that of an overfall spillway described in Section 7.5. If, however, the water that flows into the reservoir exceeds the capacity of the overfall, the water level at the crest will continuously rise until it reaches and passes the level of the crown, C. At this point the conduit is primed and siphon action begins. Theoretically, the discharge head is increased by the amount equal to $(H - H_a)$ (Figure 7.16), and the discharge rate can thus be materially increased. The large head allows rapid discharge of the excess water until it drops to the near entrance elevation.

The portion of the spillway conduit rising above the hydraulic gradient line (HGL) is under negative pressure. Since the hydraulic gradient line represents the zero atmospheric pressure, the vertical distance measured between the hydraulic gradient line and the conduit immediately above the HGL indicates the negative pressure head, $-P/\gamma$, at the location. The crown of a siphon is the highest point in the conduit, and, hence, it is subjected to the maximum negative pressure. The maximum negative pressure at a spillway

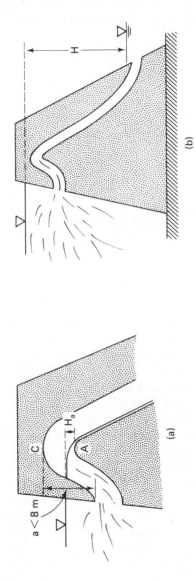

Figure 7.16. Schematical representation of a siphon spillway. (a) Beginning stage. (b) Siphoning stage.

crown must not be allowed to decrease below the vapor pressure of water at the temperature measured.

(If the negative pressure at any section in the conduit drops below the water vapor pressure, the liquid vaporizes, and vast numbers of small vapor cavities form. These vapor bubbles are carried down the conduit with the flow. When a cavity bubble reaches the region of higher pressure, the vapor condenses again into liquid and sudden collapse takes place. As the bubble collapses, water surrounding the bubble cavity rushes in at great speed to fill the cavity. At that instant the water meets in the cavity with great momentum, which is converted into damaging pressure.)

Since, under ordinary conditions, the atmospheric pressure is equivalent to 10.3 m height of water column, the maximum distance between the crown (highest point in the siphon) and the water surface elevation in the reservoir is limited to approximately 8 m. This difference allows us to include the vapor pressure head, the velocity head, and the head losses between the reservoir and the crown.

Example 7.3

The rectangular siphon as shown in Fig. 7.17 has a constant cross section of 1 m × 1 m and is 40 m long. The distance between the entrance and the crown is 10 m, the friction factor $f = 0.025$, the inlet coefficient $= 0.1$, the bend loss coefficient $= 0.8$ at the crown, and the exit loss coefficient $= 1.0$. Determine the discharge and the pressure head at the crown section.

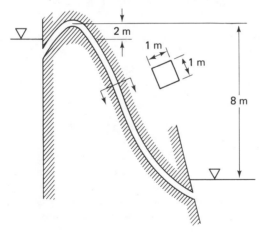

Figure 7.17

Solution

The energy relationship between point 1 (the upstream reservoir) and point 2 (the outlet) may be written as

$$\frac{V_1^2}{2g} + \frac{P_1}{\gamma} + z_1 = \frac{V_2^2}{2g} + \frac{P_2}{\gamma} + z_2 + 0.1\frac{V_2^2}{2g} + 0.8\frac{V_2^2}{2g} + 0.025\left(\frac{L}{D}\right)\frac{V_2^2}{2g}$$

Here, $V_1 = 0$ and $P_1/\gamma = P_2/\gamma = 0$; $z_1 = 6$ m and $z_2 = 0$. The above equation may thus be simplified, to

$$6 = \left[1 + 0.1 + 0.8 + 0.025\left(\frac{40}{1}\right)\right]\frac{V_2^2}{2g}$$

$$V_2 = 6.37 \text{ m/s}$$

Hence, the discharge Q is

$$Q = AV = (1)^2(6.37) = 6.37 \text{ m}^3/\text{s}$$

The energy relationship between point 1 (the reservoir) and point c (the crown) may be expressed as

$$\frac{V_1^2}{2g} + \frac{P_1}{\gamma} + 6 = \frac{V_c^2}{2g} + \frac{P_c}{\gamma} + 8 + 0.1\frac{V_c^2}{2g} + 0.025\left(\frac{10}{1}\right)\frac{V_c^2}{2g} + 0.8\frac{V_c^2}{2g}$$

This equation may be simplified since, $V_1 = 0$, $P_1/\gamma = 0$, and $V_c = V_2 = 6.37$ m/s.

$$6 = \frac{(6.37)^2}{2 \cdot 9.81}(1 + 0.1 + 0.8 + 0.25) + \frac{P_c}{\gamma} + 8$$

Therefore, the pressure head at the crown section is P_c/γ:

$$\frac{P_c}{\gamma} = -6.45 \text{ m}$$

PROBLEMS

7.6.1. Water is discharged at the rate of 0.1 m³/s in a siphon 50 m long. The crown is located 1.2 m above the reservoir water elevation and 10 m away from the entrance. If the siphon diameter is 20 cm, the friction factor is 0.021, and the head loss coefficients are the same as in Example 7.3, determine the elevation difference between the crown section and the downstream reservoir. What is the pressure at the crown?

7.6.2. A siphon spillway has a throat section of 1 m in height and a uniform width of 2 m. The siphon discharges at elevation 100 m. The siphon is primed when the headwater reaches the elevation of 110 m. The total energy loss in the conduit, exclusive of outlet velocity head, is 3.2 m. Determine the discharge.

7.7 CULVERTS

Culverts are built at the points of lowest valley to pass water across the embankments of highways or railroads. The objective of the hydraulic design of culverts is to determine the most economic dimension that can provide the passage of a designed discharge without exceeding the allowable headwater elevation. The major components of a culvert normally include the inlet, the pipe barrel, the outlet, and an outlet energy dissipator, if necessary.

Inlet structures protect embankments from erosion and improve the hydraulic conditions of culverts. Outlet structures are designed to protect culvert outlets from scouring. Although simple in appearance, the hydraulics of culverts may involve both that of the pressure pipe and the open channel. Under various discharge conditions, the hydraulic operation of culverts may be classified into four categories: (1) submerged inlet and outlet, (2) submerged inlet with full flow but free discharge at the outlet, (3) submerged inlet with partially full pipe, and (4) unsubmerged inlet. These categories are shown in Figure 7.18. The hydraulic conditions are discussed below.

1. Submergence of the culvert outlet may result from inadequate drainage downstream. In this case, the culvert discharge is determined primarily by the tail water elevation (TW) and the head loss of the culvert, regardless of the culvert slope. The culvert flow can be treated as a pressure pipe, and the head loss of the culvert, h_L, is the sum of an entrance loss, h_{ent}, friction loss, h_f, and the velocity head in the barrel.

$$h_L = h_{ent} + h_f + \frac{V^2}{2g}$$

Substituting Eqs. (3.18) and (3.16) for h_{ent} and h_f, respectively, we have

$$h_L = K_{ent}\left(\frac{V^2}{2g}\right) + \frac{n^2 V^2 L}{R_H^{4/3}} + \frac{V^2}{2g} \qquad (7.15)$$

The entrance coefficient, K_{ent}, is approximately 0.5 for a square-edged entrance and approximately 0.1 for a well-rounded entrance. Common values used for the Manning's roughness coefficient are $n = 0.012$ for concrete pipe and $n = 0.024$ for corrugated steel pipe. Equation (7.15) may be rearranged to express the direct relationship between the discharge and the dimensions of the culvert at any given elevation difference, h_L, and between tail water and head water.

$$h_L = \left[K_{ent} + \left(\frac{n^2 L}{R_H^{4/3}}\right)(2g) + 1\right]\frac{8Q^2}{\pi^2 g D^4} \qquad (7.16)$$

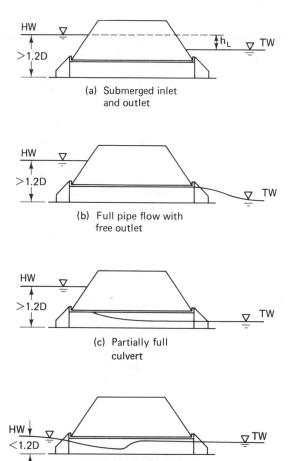

Figure 7.18. Hydraulic operation of culverts. (a) Submerged inlet and outlet. (b) Full pipe flow with free outlet. (c) Partially full culvert. (d) Unsubmerged inlet and outlet.

where Q is the discharge, D is the diameter, and R_H is the hydraulic radius ($R_H = D/4$) of the culvert barrel.

2. If the discharge carried in a culvert has a normal depth larger than the barrel height, the culvert will flow full even if the tail water level drops below that of the outlet. In this case, the discharge is controlled by the head loss, and the level of the head water (HW). The hydraulics are the same as discussed above.

3. If the normal depth is less than the barrel height, with the inlet submerged and free discharge at the outlet, a partially full pipe flow con-

dition will normally result, as illustrated in Figure 7.18(c). The culvert discharge is controlled by the entrance condition, and the flow is said to be under *entrance* control. The discharge can be calculated by

$$Q = C_d A \sqrt{2gh} \tag{7.17}$$

where h is the hydrostatic head above the center of the orifice and A is the cross-sectional area. C_d is the coefficient of discharge; common values used in practice are $C_d = 0.62$ for square-edged entrance and $C_d = 1.0$ for well-rounded entrance.

4. When the hydrostatic head at the entrance is less than $1.2D$, air will break into the barrel and the culvert will flow under no pressure. In this case, the culvert slope and the barrel wall friction determine the flow condition in the culvert for open channel flow. Due to a sudden reduction of the water area at the entrance, the flow usually enters the culvert in a supercritical condition. The critical depth takes place at the entrance of the barrel. The friction of the barrel wall gradually dissipates the energy. If the rate of dissipation is higher than the flow could gain from the barrel slope, the depth of the flowing water will increase in the downstream direction. Depending on the tail water level, the supercritical flow may convert to subcritical flow through a hydraulic jump. The flow conditions can be computed by applying the water surface profiles developed for open channels.

Example 7.4

A corrugated steel pipe is used as a culvert that must carry a flow rate of 5.3 m³/s and discharge into the air. At the entrance, the maximum available water head is 3.2 m above the bottom as shown in Figure 7.19. The culvert is 35 m long and has a square-edged entrance and slope of 0.003. Determine the diameter of the pipe.

Figure 7.19

Solution

(a) Allowing full pipe flow, the energy balance of the culvert flow may be expressed as (see Figure 7.19),

$$h_L = H - D + S_0 L = 3.2 - D + 0.003 \cdot 35$$
$$= 3.305 - D$$

where D is the diameter of the pipe. Also from Eq. (7.16), we have,

$$h_L = \left(K_{\text{ent}} + \frac{n^2 L}{R_H^{4/3}} \cdot 2g + 1\right) \frac{8Q^2}{\pi^2 g D^4}$$
$$= \left[0.5 + \frac{(0.024)^2 \cdot 35 \cdot 19.62}{(D/4)^{4/3}}\right] \frac{8 \cdot (5.3)^2}{\pi^2 \cdot 9.81 \cdot D^4}$$

Combining both equations, we get

$$D + \left(1.5 + \frac{2.51}{D^{4/3}}\right) \cdot \left(\frac{2.321}{D^4}\right) = 3.305$$

By trial, it is found that $D = 1.395$ m.

(b) If the pipe flow is partially full, then the discharge is controlled by the entrance condition only. In this case, the head h is measured above the centerline of the pipe, and we have,

$$h + \frac{D}{2} = 3.2$$

so that

$$h = 3.2 - \frac{D}{2}$$

and the orifice formula, Eq. (7.17), shall be used to compute the discharge

$$Q = 5.3 = C_D A \sqrt{2gh} = C_D \pi (D/2)^2 \sqrt{2gh}$$
$$= 0.62\pi (D/2)^2 \sqrt{2 \cdot 9.81 \cdot (3.2 - D/2)}$$

By trial, it is found that $D = 1.24$ m

Because the rate of friction dissipation is higher than the barrel slope, it is evident that the adequate solution is the one in part (a).

PROBLEMS

7.7.1. A culvert is installed on a slope of 0.05. It is designed to discharge 7.5 m³/s at the maximum head water elevation of 4.8 m. Determine the diameter of the corrugated steel pipe to be used if the length of the culvert is 12 m.

7.7.2. A rectangular concrete conduit is to be used as a culvert on a slope of 0.02. The culvert is 15 m long and has a cross section of 2 m × 2 m. If the tail water elevation is 1.8 m above the crown at the outlet, determine the head water elevation necessary to pass a 10 m³/s discharge.

7.7.3. A circular corrugated metal highway culvert is 60 m long and is laid on a slope of 0.1, with free outlet. Assume that the Manning coefficient for the corrugated metal is $n = 0.022$ and that the inlet contraction coefficient is 0.6. The culvert is designed to pass a 2.5 m³/s discharge when the head water elevation is 0.5 m above the crown at the inlet. Determine the size of the pipe to be used. Neglect the approaching velocity head.

7.8 STILLING BASINS

When the water velocity at an outlet is very high, the excessive amount of kinetic energy carried in the flow may be devastating to the outlet structure or its surroundings. A transitional energy dissipating structure is used. This is especially true at the end of a spillway where spilled water is highly accelerated so that the direct disposal in the downstream channel may produce enormous erosion. To avoid the damaging process, a transition structure commonly known as a *stilling basin* may be constructed. The water flow changes from the supercritical state to the subcritical state which contains much less damaging power. The stilling basin may be horizontal or inclined to match the slope of the channel. In any case, it should provide the obstruction (or friction force) sufficient to overcome the gravitational forces so that the flow can be decelerated until a hydraulic jump is processed.

The relation between the energy to be dissipated and the depth of the flow in the basin can be associated to the Froude number, $V/\sqrt{gD}$, as will be discussed in Section 8.3. Generally speaking, no special stilling basin is needed when the outlet flow Froude number is less than 1.7. As the Froude number increases, *energy dissipators* such as baffles, sills, and blocks may be installed along the floor of the basin to enhance the energy dissipation within the limited length of the basin. The U.S. Bureau of Reclamation has obtained a comprehensive set of curves to define the dimensions of both the stilling basin and the various types of dissipators. These curves, which are based on extensive experimental data, are shown in Figures 7.20, 7.21, and 7.22.

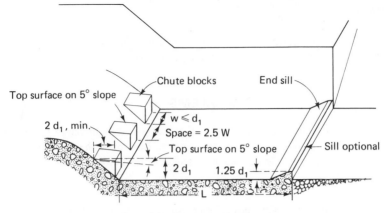

(a) Type IV basin dimensions

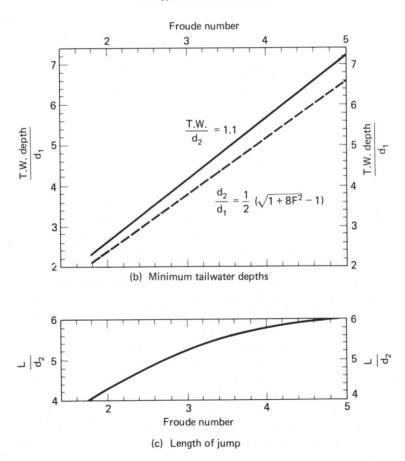

(b) Minimum tailwater depths

(c) Length of jump

Figure 7.20. Energy dissipator. The U.S. Bureau of Reclamation Type IV (for approaching Froude number between 2.5 and 4.5). (Courtesy of U.S. Bureau of Reclamation.)

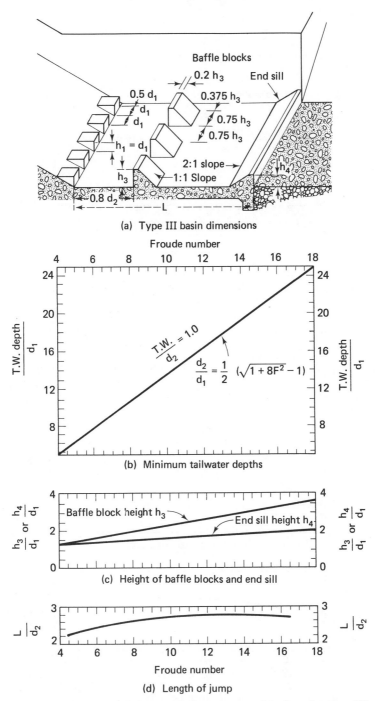

(a) Type III basin dimensions

(b) Minimum tailwater depths

(c) Height of baffle blocks and end sill

(d) Length of jump

Figure 7.21. Energy dissipator. The U.S. Bureau of Reclamation Type III (for Froude number above 4.5 and approaching velocity less than 20 m/s). (Courtesy of U.S. Bureau of Reclamation.)

Figure 7.22. Energy dissipator. The U.S. Bureau of Reclamation Type II (for Froude number above 4.5). (Courtesy of U.S. Bureau of Reclamation.)

7.9 STANDPIPES

Standpipes are used principally for alleviating the transient pressures in large pipeline systems. A pipeline conveying a large quantity of water over a large distance involves a gigantic mass of water in motion. If the main valve is suddenly closed, the water inertia creates a depression downstream of the valve, and the negative surge pressure may lead to collapse of the system. Similar situations may happen when a turbine in a hydraulic power plant starts to work or rapidly increases its power production. The increased flow rate pulls the hydraulic gradeline under the pipeline and creates a negative pressure in the system. A properly located standpipe, in this situation, will provide for an extra amount of water which can feed the pipeline during the transient period and allow the downstream flow to accelerate without causing a depression in pressure (see Figure 7.23).

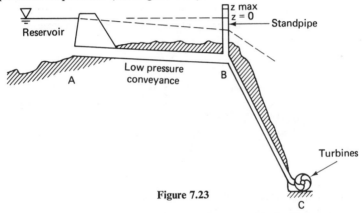

Figure 7.23

The standpipe also works as a pressure relief valve for the upstream pipe during the turbine shut-off. The water hammer effect will be diminished noticeably. The rising of water in the standpipe provides an additional hydrostatic head which decelerates the water and lowers the transient pressure in the upstream segment (AB).

REVIEWING PROBLEMS

7.1 For the concrete gravity dam shown in Figure 7.24, compute the foundation bearing pressure at the heel and toe. Assuming that the uplift force takes a triangular distribution with maximum magnitude of 1/3 of that of the hydrostatic pressure at the heel and zero at the toe. The reservoir is full to its designed elevation with 3 m of freeboard, and the masonry has specific gravity of 2.65.

7.2 A constant angle arch dam ($2\alpha = 133°34'$) is designed to span a uniform canyon 150 m wide. The height of the dam is 78 m including 3 m of freeboard. If the dam has 4 m thickness at the crest and has a symmetrical cross section that increases

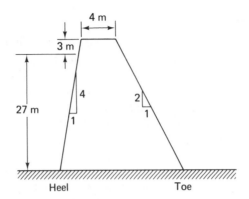

Figure 7.24

its thickness to 11.8 m at the base, determine the masonry stress in the dam, using the cylinder method (constant radii) for approximate analysis at
  (a) the crest
  (b) the midheight
  (c) the dam base.

7.3 A spillway is to be designed to carry a peak flow of 50 m³/s with the reservoir elevation 1 m above the crest of the spillway. The elevation difference between the reservoir and the tailwater is 15 m.
  (a) If an overflow spillway is used, with a crest coefficient of 2.0, determine the length of the spillway crest required to handle the discharge.
  (b) If a rectangular siphon spillway is used, with a height of 1 m and total head loss of $5V^2/2g$, determine the width of the siphon.

7.4 A horizontal rectangular stilling basin of U.S.B.R. Type III is used at the outlet of a spillway to dissipate energy. The spillway discharges 10 m³/s and has a uniform width of 10 m. At the point where the water enters the basin, the velocity is 10 m/s. Compute:
  (a) the depth of the hydraulic jump
  (b) the length of the jump
  (c) the energy loss in the jump
  (d) the efficiency of the jump defined as the ratio of specific energy after and before the hydraulic jump.

7.5 A culvert is to be designed to handle a maximum discharge of 20 m²/s across a length of 100 m. The design elevation difference between the head water at the entrance and the tailwater at the exit is 5.5 m. The design condition requires that both inlet and outlet be square-edged and flush with the walls. Assume that a concrete pipe of circular cross section is to be used which has a Manning coefficient of $n = 0.013$. Select a pipe diameter based on the slope of your choice. Neglect the approaching velocity head.

8

HYDRAULIC SIMILITUDE AND MODEL STUDIES

Use of small models for studying the prototype hydraulic design can be dated at least to Leonardo da Vinci.* But the method developed for using the results of experiments conducted on a *scaled model* to predict quantitatively the behavior of a full-size hydraulic structure (or prototype) was realized only after the turn of this century. The principle on which the model studies are based comprises the theory of *hydraulic similitude*. The analysis of the basic relationship of the various physical quantities involved in the static and dynamic behaviors of water flow in a hydraulic structure is known as *dimensional analysis*.

All important hydraulic structures are now designed and built involving certain preliminary model studies. Such studies may be conducted for any one or more of the following purposes:

1. The determination of the discharge coefficient of a large measurement structure, such as an overflow spillway or a weir.

2. The development of an effective method for energy dissipation at the outlet of a hydraulic structure.

3. The reduction of energy loss at the intake structure or at the transition section.

*Leonardo da Vinci (1452–1519), a genius, Renaissance scientist, engineer, architect, painter, sculptor, and musician.

4. The development of an efficient, economic spillway or other type of flood-releasing structure for a reservoir.
5. The determination of the average time of travel in a temperature control structure, for example, in a cooling pond in a power plant.
6. The determination of the best cross section, location, and dimensions of the various components of the structure, such as the breakwater, the docks, and the locks, etc, in harbor and waterway design.
7. The determination of the dynamic behaviors of the floating, semi-immersible, and floor-installed structures in transportation and installation of offshore structures.

River models have also been extensively used in hydraulic engineering to determine:

1. The pattern a flood wave travels through a river channel.
2. The effect of artificial structures, such as bends, levees, dikes, jetties, and training walls, on the sedimentation movements in the channel reach, as well as in the upstream and downstream channels.
3. The direction and force of currents in the channel or harbor and their effects on navigation and marine life.

8.1   PRINCIPLES OF HYDRAULIC SIMILITUDE

Similarity between hydraulic models and prototypes may be achieved in three basic forms: geometric similarity, kinematic similarity, and dynamic similarity.

Geometric similarity implies similarity of form. The model is a geometric reduction of the prototype and is accomplished by maintaining a fixed ratio for all homologous lengths between the model and the prototype.

The physical quantities involved in geometric similarity are length (L), area (A), and volume (Vol). The ratio of homologous lengths in prototype and model is a constant and can be expressed as

$$\frac{L_p}{L_m} = L_r \tag{8.1}$$

An area (A) is the product of two homologous lengths; hence, the ratio of the homologous area is also a constant and can be expressed as

$$\frac{A_p}{A_m} = \frac{L_p^2}{L_m^2} = L_r^2 \tag{8.2}$$

A volume (Vol) is the product of three homologous lengths; the ratio of the homologous volume can be expressed as

$$\frac{\text{Vol}_p}{\text{Vol}_m} = \frac{L_p^3}{L_m^3} = L_r^3 \tag{8.3}$$

Example 8.1

A geometrically similar open channel model is constructed with a 5:1 scale. If the model measures a discharge of 0.2 m³/s, determine the corresponding discharge in the prototype.

Solution

The velocity ratio between the prototype and the model is

$$\frac{V_p}{V_m} = \frac{\dfrac{L_p}{T}}{\dfrac{L_m}{T}} = \frac{L_p}{L_m} = L_r = 5$$

Since in geometric similarity, time in model and prototype remains unscaled. The area ratio between the prototype and the model is

$$\frac{A_p}{A_m} = \frac{L_p^2}{L_m^2} = L_r^2 = 25$$

Accordingly, the discharge ratio is

$$\frac{Q_p}{Q_m} = \frac{A_p V_p}{A_m V_m} = (25)(5) = 125$$

Thus, the corresponding discharge in the prototype is

$$Q_p = 125 Q_m = 125 \cdot 0.2 = 25 \text{ m}^3/\text{s}$$

Kinematic similarity implies similarity in motion. Kinematic similarity between a model and the prototype is attained if the homologous moving particles have the same velocity ratio along geometrically similar paths. The kinematic similarity involves the scale of time as well as length. The ratio of times required for homologous particles to travel homologous distances in a model and its prototype is

$$\frac{T_p}{T_m} = T_r \tag{8.4}$$

The velocity V is defined in terms of distance per unit time; thus, the ratio of velocities can be expressed as

$$\frac{V_p}{V_m} = \frac{\dfrac{L_p}{T_p}}{\dfrac{L_m}{T_m}} = \frac{\dfrac{L_p}{L_m}}{\dfrac{T_p}{T_m}} = \frac{L_r}{T_r} \tag{8.5}$$

The acceleration, a, is defined in terms of length per unit time square; thus, the ratio of homologous acceleration is

$$\frac{a_p}{a_m} = \frac{\dfrac{L_p}{T_p^2}}{\dfrac{L_m}{T_m^2}} = \frac{\dfrac{L_p}{L_m}}{\dfrac{T_p^2}{T_m^2}} = \frac{L_r}{T_r^2} \tag{8.6}$$

The discharge, Q, is expressed in terms of volume per unit time; thus,

$$\frac{Q_p}{Q_m} = \frac{\dfrac{L_p^3}{T_p}}{\dfrac{L_m^3}{T_m}} = \frac{\dfrac{L_p^3}{L_m^3}}{\dfrac{T_p}{T_m}} = \frac{L_r^3}{T_r} \tag{8.7}$$

Kinematic models constructed for hydraulic machinery may frequently involve angular displacement, θ, expressed in radians, which is equal to the tangential displacement, L, divided by the length of radius, R, of the curve at the point of tangency. The ratio of angular displacements may be expressed as

$$\frac{\theta_p}{\theta_m} = \frac{\dfrac{L_p}{R_p}}{\dfrac{L_m}{R_m}} = \frac{\dfrac{L_p}{L_m}}{\dfrac{R_p}{R_m}} = \frac{\dfrac{L_p}{L_m}}{\dfrac{L_p}{L_m}} = 1 \tag{8.8}$$

The angular velocity, N, in revolutions per minute, is defined as angular displacement per unit time; thus, the ratio

$$\frac{N_p}{N_m} = \frac{\dfrac{\theta_p}{T_p}}{\dfrac{\theta_m}{T_m}} = \frac{\dfrac{\theta_p}{\theta_m}}{\dfrac{T_p}{T_m}} = \frac{1}{T_r} \tag{8.9}$$

The angular acceleration, α, is defined as angular displacement per unit time square,

$$\frac{\alpha_p}{\alpha_m} = \frac{\frac{\theta_p}{T_p^2}}{\frac{\theta_m}{T_m^2}} = \frac{\frac{\theta_p}{T_p^2}}{\frac{\theta_m}{T_m^2}} = \frac{1}{T_r^2} \tag{8.10}$$

Example 8.2

A 10:1 scale model is constructed to study the flow motions in a cooling pond. If the designed discharge from the power plant is 200 m³/s and the model can accommodate a maximum flow rate of 0.1 m³/s, determine the time ratio.

Solution

The length ratio between the prototype and the model is

$$L_r = \frac{L_p}{L_m} = 10$$

and the discharge ratio is $Q_r = \dfrac{200}{0.1} = 2000$, and

$$Q_r = \frac{Q_p}{Q_m} = \frac{\frac{L_p^3}{T_p}}{\frac{L_m^3}{T_m}} = \left(\frac{L_p}{L_m}\right)^3 \cdot \left(\frac{T_m}{T_p}\right) = L_r^3 \cdot T_r^{-1}$$

Substituting the length ratio into the dischange ratio gives the time ratio:

$$T_r = \frac{T_p}{T_m} = \frac{L_r^3}{Q_r} = \frac{(10)^3}{2000} = 0.5$$

and

$$T_m = 2T_p$$

A unit time period measured in the model is equivalent to two periods of time in the prototype pond.

Dynamic similarity implies similarity in forces involved in motion. Dynamic similarity between a model and its prototype is attained if the ratio of homologous forces in the model and prototype is kept at a constant value, or

$$\frac{F_p}{F_m} = F_r \tag{8.11}$$

Since force is equal to mass, M, times acceleration, a, and since mass is equal to density, ρ, times volume, Vol, the force ratio may be expressed as

$$\frac{F_p}{F_m} = \frac{M_p a_p}{M_m a_m} = \frac{\rho_p \, \text{Vol}_p \, a_p}{\rho_m \, \text{Vol}_m \, a_m}$$

$$= \frac{\rho_p}{\rho_m} \cdot \frac{L_p^3}{L_m^3} \cdot \frac{\dfrac{L_p}{T_p^2}}{\dfrac{L_m}{T_m^2}} = \frac{\rho_p}{\rho_m} \cdot \frac{L_p^4}{L_m^4} \frac{\dfrac{1}{T_p^2}}{\dfrac{1}{T_m^2}}$$

$$= \rho_r L_r^4 T_r^{-2} \tag{8.12}$$

and the ratio of homologous masses, force divided by acceleration, may be expressed as

$$\frac{M_p}{M_m} = \frac{\dfrac{F_p}{a_p}}{\dfrac{F_m}{a_m}} = \frac{\dfrac{F_p}{F_m}}{\dfrac{a_p}{a_m}} = F_r T_r^2 L_r^{-1} \tag{8.13}$$

Since work is equal to force times distance, the ratio of homologous work in dynamic similarity is

$$\frac{\bar{W}_p}{\bar{W}_m} = \frac{F_p L_p}{F_m L_m} = F_r L_r \tag{8.14}$$

Power is the time rate of doing work; thus, the ratio of homologous powers in the model and prototype is

$$\frac{P_p}{P_m} = \frac{\dfrac{\bar{W}_p}{T_p}}{\dfrac{\bar{W}_m}{T_m}} = \frac{\bar{W}_p}{\bar{W}_m} \cdot \frac{1}{\dfrac{T_p}{T_m}} = \frac{F_r L_r}{T_r} \tag{8.15}$$

Example 8.3

A 59,680-W (80-hp) pump is used to power a water supply system. The model constructed to study the system has a 8:1 scale. If the velocity ratio is 2:1, determine the power needed for the model pump.

Solution

By substituting the length ratio into the velocity ratio, the time ratio is obtained:

$$V_r = \frac{L_r}{T_r} = 2; \qquad L_r = 8$$

$$T_r = \frac{L_r}{2} = \frac{8}{2} = 4$$

The same fluid is used in the model and the prototype, unless otherwise specified; then $\rho_r = 1$, and the force ratio can be calculated, from Eq. (8.12):

$$F_r = \rho_r L_r^4 T_r^{-2} = \frac{(1)(8)^4}{(4)^2} = 256$$

The power ratio is, from Eq. (8.15):

$$P_r = \frac{F_r L_r}{T_r} = \frac{(256)(8)}{(4)} = 512$$

And the power required for the model pump is:

$$P_m = \frac{P_p}{P_r} = \frac{59,680}{512} = 116.6 \text{ W} \equiv 0.156 \text{ hp}$$

Example 8.4

The model designed to study the prototype of a hydraulic machine must (a) be geometrically similar, (b) have the same discharge coefficient defined as $Q/A\sqrt{2gH}$, and (c) have the same ratio of peripheral speed to the water discharge velocity, $\omega D/(QA)$. Determine the scale ratios in terms of discharge Q, head H, diameter D, and rotational angular velocity ω.

Solution

It is important to recognize that the energy head H, although expressed in units of length, is not necessarily modeled as a linear length dimension. To have the same ratio of peripheral speed to the water discharge velocity, we have,

$$\frac{\omega_p D_p}{Q_p/A_p} = \frac{\omega_m D_m}{Q_m/A_m}$$

or,

$$\frac{\omega_r D_r A_r}{Q_r} = \frac{T^{-1} L_r L_r^2}{L_r^3 T_r^{-1}} = 1$$

To have the same discharge coefficient, we have,

$$\frac{Q_p/(A_p\sqrt{2gH_p})}{Q_m/(A_m\sqrt{2gH_m})} = \frac{Q_r}{A_r\sqrt{(gH)_r}} = 1$$

or,

$$\frac{L_r^3 T_r^{-1}}{L_r^2 (gH)_r^{1/2}} = 1$$

From which, we get,

$$(gH)_r = \frac{L_r^2}{T_r^2}$$

Since the gravitational acceleration, g, is the same for model and prototype, we may write,

$$H_r = L_r^2 T_r^{-2}$$

The other ratio asked are:

Discharge ratio, $Q_r = L_r^3 T_r^{-1}$
Diameter ratio, $D_r = L_r$, and
Angular speed ratio, $\omega_r = T_r^{-1}$

PROBLEMS

8.1.1. A 1-m long, 1 : 30 model is used to study the wave force on a prototype of a sea wall structure. If the total wave force measured on the model is 2.27 N and the velocity scale is 1 : 10, determine the force per unit length of the prototype.

8.1.2. The moment exerted on a gate structure is studied in a laboratory water tank with a 1 : 125 scale model. If the moment measured on the model is 1.5 N•m on the 1-m long gate arm, determine the moment exerted on the prototype.

8.1.3. A 1:100 scale model is constructed to study a gate prototype which is designed to drain a reservoir. If the model reservoir is drained in 5.2 min, how long should it take to drain the prototype?

8.1.4. An overflow spillway is designed to have a 100-m long crest to carry the discharge of 1150 m³/s under a permitted maximum head of 3 m. The operation of the prototype spillway is studied by a 1 : 50 scale model in a hydraulic laboratory.
(a) Determine the model discharge.
(b) The model velocity measured at the end (toe) of the spillway is 3 m/s. Determine the corresponding velocity in the prototype.
(c) What are the model and prototype Froude numbers at the toe?
(d) If the force on a bucket energy dissipator at the toe of the spillway is measured to be 37.5 N, determine the corresponding force on the prototype.

8.1.5. If a 1 : 5 scale model, at 1200 rpm, is used to study the prototype of a centrifugal pump that produces 1 m³/s at 30 m head when rotating at 400 rpm, determine the model discharge and head.

8.2 PHENOMENA GOVERNED BY VISCOUS FORCE—REYNOLDS NUMBER LAW

Water in motion always involves inertial forces. When the inertial forces and the viscous forces can be considered to be the only forces that govern the motion, the ratio of these forces acting on homologous particles in a model and its prototype is defined by the Reynolds law

$$N_R = \frac{\text{inertial force}}{\text{viscous force}} \tag{8.16}$$

The inertial force defined by Newton's second law of motion, $F = ma$, can be expressed by the ratio in Eq. (8.12):

$$F_r = M_r \frac{L_r}{T_r^2} = \rho_r L_r^4 T_r^{-2} \tag{8.12}$$

The viscous force defined by Newton's law of viscosity,

$$F = \mu\left(\frac{dV}{dL}\right) \cdot A,$$

may be expressed by

$$F_r = \frac{\mu_p \left(\frac{dV}{dL}\right)_p A_p}{\mu_m \left(\frac{dV}{dL}\right)_m A_m} = \mu_r L_r^2 T_r^{-1} \tag{8.17}$$

where μ is the viscosity and V denotes the velocity.

Equating values of F_r from Eqs. (8.12) and (8.17), we get

$$\rho_r L_r^4 T_r^{-2} = \mu_r L_r^2 T_r^{-1}$$

from which

$$\frac{\rho_r L_r^4 T_r^{-2}}{\mu_r L_r^2 T_r^{-1}} = \frac{\rho_r L_r^2}{\mu_r T_r} = \frac{\rho_r L_r V_r}{\mu_r} = 1 \tag{8.18}$$

Rearranging the above equation, we may write

$$\frac{\frac{\rho_p L_p V_p}{\mu_p}}{\frac{\rho_m L_m V_m}{\mu_m}} = 1$$

or

$$\frac{\rho_p L_p V_p}{\mu_p} = \frac{\rho_m L_m V_m}{\mu_m} = N_R \tag{8.19}$$

The above equation states that when the inertial force and the viscous force are considered to be the only forces governing the motion of the water, the Reynolds number of the model and the prototype must be kept at the same value.

If the same fluid is used in both the model and the prototype, the scale ratios for many physical quantities can be derived based on the Reynolds number law. These quantities are listed in Table 8.1.

TABLE 8.1 Scale Ratios for Reynolds Number Law (for Water in Both Model and Prototype, $\rho_r = 1$, $\mu_r = 1$)

Geometric Similarity		Kinematic Similarity		Dynamic Similarity	
Length	L_r	Time	L_r^2	Force	1
Area	L_r^2	Velocity	L_r^{-1}	Mass	L_r^3
Volume	L_r^3	Acceleration	L_r^{-3}	Work	L_r
		Discharge	L_r	Power	L_r^3
		Angular Velocity	L_r^{-2}		
		Angular Acceleration	L_r^{-4}		

Example 8.5

In order to study a transient process, a model is constructed at a 10:1 scale. Water is used in the prototype, and it is known that viscous forces are the dominant ones. Compare the time and force ratios, if the model uses (a) water and (b) oil 5 times more viscous than water, with $\rho_{oil} = 0.8\rho_w$.

Solution

(a) From Table 8.1

$$T_r = L_r^2 = (10)^2 = 100$$
$$V_r = L_r^{-1} = (10)^{-1} = 0.1$$
$$F_r = 1$$

(b) From the Reynolds number law,

$$\frac{\rho_p L_p V_p}{\mu_p} = \frac{\rho_m L_m V_m}{\mu_m}$$

we have

$$\frac{\rho_r L_r V_r}{\mu_r} = 1$$

Since the ratios of viscosity and density are, respectively,

$$\mu_r = \frac{\mu_p}{\mu_m} = \frac{\mu_{\text{water}}}{\mu_{\text{oil}}} = \frac{\mu_{\text{water}}}{5\mu_{\text{water}}} = 0.2$$

$$\rho_r = \frac{\rho_p}{\rho_m} = \frac{\rho_{\text{water}}}{\rho_{\text{oil}}} = \frac{\rho_{\text{water}}}{0.8\rho_{\text{water}}} = 1.25$$

From the Reynolds number law

$$V_r = \frac{\mu_r}{\rho_r L_r} = \frac{(0.2)}{(1.25)(10)} = 0.016$$

The time ratio is

$$T_r = \frac{L_r}{V_r} = \frac{\rho_r L_r^2}{\mu_r} = \frac{(1.25)(10)^2}{(0.2)} = 625$$

The force ratio, then, is

$$F_r = \frac{\rho_r L_r^4}{T_r^2} = \frac{(\rho_r L_r^4)}{\left(\frac{\rho_r^2 L_r^4}{\mu_r^2}\right)} = \frac{\mu_r^2}{\rho_r} = \frac{(0.2)^2}{1.25} = 0.032$$

This example demonstrates the importance of selecting the model fluid. The properties of the liquid used in the model, especially the viscosity, greatly affect the performance in the Reynolds number models.

PROBLEMS

8.2.1. A Reynolds number scale model is used to study the operation of a prototype hydraulic device. The model is built on a 1: 5 scale and uses water at 20°C. The prototype discharges 11.5 m³/s of water at 90°C temperature. Determine the model discharge.

8.2.2. The moment exerted on a ship's rudder is studied with a 1: 20 scale model in a water tunnel using the same temperature as the river water. If the torque measured on the model is 10 N·m for a water tunnel velocity of 20 m/s, determine the corresponding torque and speed for the prototype.

8.2.3. A 1: 10 scale model of a water supply piping system is to be tested at 20°C to determine the total head loss in the prototype which carries water at 85°C. The prototype is designed to carry 5.0 m³/s discharge with 1-m diameter pipes. Determine the model discharge and model velocity. Discuss how losses determined from the model are converted to the prototype losses.

8.2.4. A submerged vehicle moves at 5 m/s in the ocean. At what theoretical speed must a 1:10 model be towed for there to be dynamic similarity between the model and the prototype? Assume that the sea water and towing tank water are the same.

8.3 PHENOMENA GOVERNED BY GRAVITY FORCE—FROUDE NUMBER LAW

When inertial force and gravity force are considered to be the only dominating forces in the fluid motions, the ratio of the inertial forces acting on the homologous elements of the fluid in the model and prototype can be defined by Eq. (8.12):

$$\frac{F_P}{F_m} = \rho_r L_r^4 T_r^{-2} \tag{8.12}$$

and the ratio of gravity forces, which is determined by the weight of the homologous fluid elements involved,

$$\frac{F_P}{F_m} = \frac{M_P g_P}{M_m g_m} = \frac{\rho_P g_P L_P^3}{\rho_m g_m L_m^3} = \rho_r g_r L_r^3 \tag{8.20}$$

Equating the values from Eqs. (8.12) and (8.20), we get

$$\rho_r L_r^4 T_r^{-2} = \rho_r g_r L_r^3$$

Rearranging, we get

$$g_r L_r = \frac{L_r^2}{T_r^2} = V_r^2$$

or

$$\frac{V_r}{g_r^{1/2} L_r^{1/2}} = 1 \tag{8.21}$$

From which

$$\frac{\dfrac{V_P}{g_P^{1/2} L_P^{1/2}}}{\dfrac{V_m}{g_m^{1/2} L_m^{1/2}}} = 1$$

Hence,

$$\frac{V_P}{g_P^{1/2} L_P^{1/2}} = \frac{V_m}{g_m^{1/2} L_m^{1/2}} = N_F \qquad \text{(Froude number)} \tag{8.22}$$

In other words, when the inertial force and the gravity force are considered to be the only forces that dominate the fluid motions, the Froude number of the model and the prototype should be kept at the same value.

If the same fluid is used in both the model and the prototype which are subjected to the same gravitational force field, many physical quantities can be derived based on the Froude number law. These quantities are listed in Table 8.2.

TABLE 8.2 Ratios for the Froude Number Law ($g_r = 1$, $\rho_r = 1$)

Geometric Similarity		Kinematic Similarity		Dynamic Similarity	
Length	L_r	Time	$L_r^{1/2}$	Force	L_r^3
Area	L_r^2	Velocity	$L_r^{1/2}$	Mass	L_r^3
Volume	L_r^3	Acceleration	1	Work	L_r^4
		Discharge	$L_r^{5/2}$	Power	$L_r^{7/2}$
		Angular Velocity	$L_r^{-1/2}$		
		Angular Acceleration	L_r^{-1}		

Example 8.6

An open channel model 30 m long is built to satisfy the Froude number law. What is the flow in the model for a prototype flood of 700 m³/s if the scale used is 20:1? Determine also the force ratio.

Solution

From Table 8.2, the discharge ratio is

$$Q_r = L_r^{5/2} = (20)^{2.5} = 1789$$

Thus, the model flow should be

$$Q_m = \frac{Q_P}{Q_r} = \frac{700 \text{ m}^3/\text{s}}{1789} = 0.391 \text{ m}^3/\text{s} = 391 \text{ l/s}$$

The force ratio is

$$F_r = \frac{F_P}{F_m} = L_r^3 = (20)^3 = 8000$$

PROBLEMS

8.3.1. An overflow spillway with a 300-m crest is designed to discharge 3600 m³/s. A 1:20 model of the cross section of the dam is built in the laboratory flume 1 m wide. Calculate the required laboratory flow rate. Neglect viscosity and surface tension effects.

8.3.2. If a 1:1000 scale tidal basin model is used to study the operation of a prototype satisfying the Froude number law, what length of time in the model represents the period of one day in the prototype?

8.3.3. A ship 100 m long designed to travel at a top speed of 1 m/s is to be studied in a towing tank with a 1:50 scale model. Determine what speed the model must be towed for (a) the Reynolds number law and (b) the Froude number law.

8.4 PHENOMENA GOVERNED BY SURFACE TENSION—WEBER NUMBER LAW

Surface tension is a measure of energy level on the surface of a liquid body. In hydraulic engineering practice the force becomes of primary importance in the study of the motion of small surface waves, or control of evaporation from a large body of water, such as a water storage tank or reservoir.

Surface tension, denoted by σ, is measured in terms of force per unit length. Hence, the force is $F = \sigma L$. The ratio of analogous surface tension forces in prototype and in model is

$$F_r = \frac{F_p}{F_m} = \frac{\sigma_p L_p}{\sigma_m L_m} = \sigma_r L_r \tag{8.23}$$

Equating the surface tension force ratio to the inertial force ratio [Eq. (8.12)] gives

$$\sigma_r L_r = \rho_r \frac{L_r^4}{T_r^2}$$

Rearranging gives

$$T_r = \left(\frac{\rho_r}{\sigma_r}\right)^{1/2} L_r^{3/2} \tag{8.24}$$

By substituting for T_r the basic relationship of $V_r = L_r/T_r$, the above equation may be rearranged to give

$$V_r = \frac{L_r}{\left(\frac{\rho_r}{\sigma_r}\right)^{1/2} L_r^{3/2}} = \left(\frac{\sigma_r}{\rho_r L_r}\right)^{1/2}$$

or

$$\frac{\rho_r V_r^2 L_r}{\sigma_r} = 1 \tag{8.25}$$

Hence,

$$\frac{\rho_p V_p^2 L_p}{\sigma_p} = \frac{\rho_m V_m^2 L_m}{\sigma_m} = N_w \qquad \text{(Weber number)} \tag{8.26}$$

In other words, the Weber number must be kept at the same value in the model and in the prototype for studying phenomena governed by surface tension force. If the same liquid is used in both model and prototype, then $\rho_r = 1.0$, and $\sigma_r = 1.0$, and the above relationship can be simplified to

$$V_r^2 L_r = 1$$

or

$$V_r = L_r^{1/2} \tag{8.27}$$

Since $V_r = L_r/T_r$, we may also write

$$\frac{L_r}{T_r} = L_r^{1/2}$$

Thus,

$$T_r = L_r^{3/2} \tag{8.28}$$

8.5 PHENOMENA GOVERNED BY BOTH GRAVITY AND VISCOUS FORCES

In the case of motion of surface vessels through water, or the propagation of shallow water waves in open channels, both gravity and viscous forces may be important. The study of these phenomena requires that both the Froude number and Reynolds number laws be satisfied simultaneously. That is,

$$\frac{\rho_r L_r V_r}{\mu_r} = \frac{V_r}{(g_r L_r)^{1/2}}$$

Assuming that both the model and the prototype are affected by the earth gravitational field, $g_r = 1$ and $v = \mu/\rho$, the above relationship may be simplified to

$$v_r = L_r^{3/2} \tag{8.29}$$

This requirement can be met only by choosing a special model fluid with a kinematic viscosity ratio to water equal to the three-half power of the scale ratio. In general, this requirement is difficult to meet. For example, a 1:10 scale model would require that the model fluid have a kinematic viscosity of 30 times less than that of water, which is obviously impossible.

Two expedients, however, may be available depending on the relative importance of the two forces in the particular phenomenon. In the case of ship resistance, the ship model may be built according to the Reynolds model law and may operate in a towing tank in accordance with the Froude number law. In the case of shallow water waves in open channels, empirical relationships such as the Manning's formula [Eq. (3.16)] may be used as an auxiliary condition for the wave measurements, according to the Froude number law.

8.6 MODELS FOR FLOATING AND SUBMERGED BODIES

Model studies for floating and submerged bodies are performed to obtain information on (1) the friction drag along the boundary of the moving vessel, (2) the form drag resulting from flow separation from the vessel boundary due to the boundary shape, (3) the force expended in generation of gravity waves, and (4) the stability of the body in withstanding the water waves and the wave forces on the body.

The first two forces are strictly viscous phenomena, and therefore the models should be designed according to the Reynolds number law. The third force is a gravity force governed phenomenon and, hence, must be analyzed by applying the Froude number law. All three measurements may be performed simultaneously in a towing water tank. In analyzing the data, however, the friction forces and the form drag forces are first computed from the measurements by using known formulas and drag coefficients. The remaining force measured in towing the vessel through the water surface is the force expended in generation of the gravity waves (wave resistance), and it is scaled up to the prototype values by the Froude number law.

The analysis procedure is demonstrated in Example (8.7). For subsurface vessels, such as submarines, the effect of surface waves on the vessels may be neglected. Hence, the Froude number model is not needed. To study the stability and wave force on stationary offshore structures, the effect of inertial force must be taken into consideration. The inertial force, defined as $F_i = M' \cdot a$, can be calculated directly from the prototype dimensions. Here, M' is the mass of water displaced by the portion of the structure immersed below the waterline (also known as the *virtual mass*), and a is the acceleration of the water mass.

Example 8.7

A ship model with a maximum cross-sectional area of 0.78 m² immersed below the waterline, has a characteristic length of 0.9 m. The model is towed in a wave tank at the speed of 0.5 m/s. For the particular shape of the vessel, it is found that the drag coefficient can be approximated by $C_D = (0.06/N_{Re}^{0.25})$ for $10^4 \leq N_{Re} \leq 10^6$, and $C_D = 0.0018$ for $N_{Re} > 10^6$. The Froude number law is applied for the 1 : 50 model. During the experiment, a total force of 0.40 N is measured. Determine the total resistance force on the prototype vessel.

Solution

Based on the Froude number law, we may determine the velocity ratio,

$$V_r = L_r^{1/2} = (50)^{1/2} = 7.07$$

Hence, the corresponding velocity of the vessel is:

$$V_p = V_r \cdot V_m = 0.5 \cdot 7.07 = 3.54 \text{ m/s}$$

The Reynolds number for the model is

$$N_{Re} = \frac{V_m \cdot L_m}{\nu} = \frac{0.5 \cdot 0.9}{1.003 \cdot 10^{-6}} = 4.49 \cdot 10^5$$

and the drag coefficient for the model is

$$C_{D_m} = \frac{0.06}{(4.49 \cdot 10^5)^{1/4}} \ 0.0023$$

The drag force on a vessel is defined as $D = C_D(\frac{1}{2}\rho A \cdot V^2)$, where ρ is the water density, and A is the projected area of the immersed portion of the vessel on a plane normal to the direction of the motion. Thus, the model drag force can be calculated as

$$D_m = C_{D_m}(\tfrac{1}{2}\rho_m A_m \cdot V_m^2) = 0.0023 \cdot \tfrac{1}{2} \cdot 1000 \cdot 0.78 \cdot 0.5^2 = 0.2243 \text{ N}$$

The model wave resistance is the difference between the measured towing force and the drag force:

$$F_w = 0.4 - 0.2243 = 0.1757 \text{ N}$$

For the prototype, the Reynolds number is

$$N_{Re} = \frac{V_p \cdot L_p}{\nu_p} = \frac{V_p \cdot L_r \cdot L_m}{\nu_p} = 1.59 \cdot 10^8$$

The drag coefficient of the prototype vessel is $C_{D_p} = 0.0018$, and the drag force is

$$D_p = C_{D_p}(\tfrac{1}{2}\rho_p A_p \cdot V_p^2) = C_{D_p}(\tfrac{1}{2}\rho_p A_m \cdot L_r^2 \cdot V_p^2) = 21937 \text{ N}$$

The wave resistance on the prototype vessel is calculated by applying Froude number law (see Table 8.2):

$$F_{w_p} = F_{w_r} \cdot F_{w_m} = L_r^3 \cdot F_{w_m} = 21963 \text{ N}$$

Hence the total resistance force on the prototype is

$$F = D_p + F_{w_p} = 21937 + 21963 = 43900 \text{ N}$$

PROBLEMS

8.6.1. A ship 100 m long moves at 1.5 m/s in freshwater at 15°C. A 1:100 scale model of the prototype ship is to be tested in a towing tank containing a liquid of specific gravity 0.9. What viscosity must this liquid have for both Reynolds and Froude number laws to be satisfied?

8.6.2. A 1:250 ship model is towed in a wave tank and a wave resistance of 10.7 N is measured. Determine the corresponding prototype wave resistance on the prototype.

8.6.3. A barge model 1 m long is tested in a towing tank at the speed of 1 m/s. Determine the prototype velocity if the prototype is 150 m in length. The model has 2 cm draft and is 10 cm wide. The drag coefficient is $C_D = 0.25$ for $N_{Re} > 5 \cdot 10^4$, and the towing force required to tow the model is 0.3 N. What force is required to tow the barge in waterways?

8.7 OPEN CHANNEL MODELS

Open channel models may be used to study either the velocity discharge-slope relationship or the effect of flow patterns on the changes in bed configuration. For the former applications, relatively long reaches of the river channel can be modeled. A special example is the U.S. Army Engineers Waterways Experiment Station in Vicksburg, Mississippi, where the Mississippi River is modeled on one site. In these applications, where the effect due to changes in bed configuration are only of secondary concern, a fixed-bed model may be used. Basically, this model is used in studying the velocity-slope relationship in a particular channel; therefore, the effect of bed roughness is important.

An empirical relation, such as the Manning equation [Eq. (3.16)], may be used to assume the similarity between the prototype and the model:

$$V_r = \frac{V_P}{V_m} = \frac{\dfrac{1}{n_p} R_{H_p}^{2/3} S_p^{1/2}}{\dfrac{1}{n_m} R_{H_m}^{2/3} S_m^{1/2}} = \frac{1}{n_r} R_{H_r}^{2/3} S_r^{1/2} \tag{8.30}$$

If the model is built with the same scale ratio for the horizontal dimensions ($\bar{X}$) and vertical dimensions ($\bar{Y}$), known as the *undistorted model*, then

$$R_{H_r} = \bar{X}_r = \bar{Y}_r = L_r, \quad \text{and} \quad S_r = 1$$

and

$$V_r = \frac{1}{n_r} L_r^{2/3} \tag{8.31}$$

Since the Manning's roughness coefficient, $n \propto R^{1/6}$ [Eq. (5.3)], we may write

$$n_r = L_r^{1/6} \tag{8.32}$$

This model may frequently result in a model velocity so small (or the model roughness will be so relatively large) that realistic measurements cannot be made; or a model water depth may be so shallow that the physical characteristics of the flow may be altered. Such situations may be resolved by using a distorted model in which the vertical scale and the horizontal scale are not of the same value; usually, a smaller vertical scale ratio, $\bar{X}_r > \bar{Y}_r$. This means that

$$S_r = \frac{S_p}{S_m} = \frac{\bar{Y}_r}{\bar{X}_r} < 1$$

Hence, $S_m > S_p$, or a larger slope for the model. The use of the Manning equation requires that the flow be fully turbulent in both model and prototype.

Open channel models involving problems of sediment transportation, erosion, or deposit require movable bed models. A movable channel bed consists of sand or other loose material that can be moved in response to the forces of the current at the channel bed. Normally, it is impractical to scale the bed material down to the model scale. A vertical scale distortion is usually employed on a movable bed model in order to provide a sufficient tractive force to induce bed material movement. Quantitative similarity is difficult to attain in movable bed models. For any sedimentation studies performed, it is important that the movable bed model be quantitatively verified by a number of field measurements.

Example 8.8

An open channel model is built to study the effects of tidal waves on sedimentation movement in a 10-km river reach (the reach meanders in an area 7 km long). The mean depth and width of the reach are 4 m and 50 m, respectively, and the discharge is 850 m³/s. The Manning's roughness coefficient $n_p = 0.035$. If the model is to be constructed in a laboratory room 18 m long, determine a convenient scale and the model discharge.

Solution

In surface wave phenomenon the gravitational forces are dominant. The Froude number law will be used for the modeling. The laboratory length will limit the horizontal scale:

$$\bar{X}_r = \frac{L_p}{L_m} = \frac{7000}{18} = 389$$

We will use $L_r = 400$ for convenience.

It is judged reasonable to use a vertical scale of $\bar{Y}_r = 80$ (enough to measure surface gradients). Recalling that the hydraulic radius is the characteristic dimension in open channel and that for a large width-to-depth ratio the hydraulic radius is roughly equal to the water depth, we can make the following approximation:

$$R_{H_r} = \bar{Y}_r = 80$$

Since

$$N_F = \frac{V_r}{g_r^{1/2} R_{H_r}^{1/2}} = 1 \qquad (8.21)$$

Then

$$V_r = R_{H_r}^{1/2} = \bar{Y}_r^{1/2} = (80)^{1/2}$$

Using Manning's formula [or Eq. (8.30)],

$$V_r = \frac{V_p}{V_m} = \frac{1}{n_r} R_{H_r}^{2/3} S_r^{1/2} \qquad S_r = \frac{\bar{Y}_r}{\bar{X}_r}$$

we have

$$n_r = \frac{R_{H_r}^{2/3} S_r^{1/2}}{V_r} = \frac{\bar{Y}_r^{2/3} \left(\frac{\bar{Y}_r}{\bar{X}_r}\right)^{1/2}}{\bar{Y}_r^{1/2}} = \frac{\bar{Y}_r^{2/3}}{\bar{X}_r^{1/2}} = \frac{(80)^{2/3}}{(400)^{1/2}} = 0.928$$

Hence,

$$n_m = \frac{n_p}{n_r} = \frac{0.035}{0.928} = 0.038$$

The discharge ratio is

$$Q_r = A_r \cdot V_r = \bar{X}_r \bar{Y}_r V_r = \bar{X}_r \bar{Y}_r^{3/2} = (400)(80)^{3/2} = 286{,}217$$

Thus, the model discharge required is

$$Q_m = \frac{Q_p}{Q_r} = \frac{850}{286{,}217} = 0.003 \text{ m}^3/\text{s} \equiv 3 \text{ l/s}$$

In order to use the Manning formula, turbulent flow must be ensured in the model. To verify the turbulent flow condition in the model, it is necessary to calculate the value of the model Reynolds number.

The horizontal prototype velocity is

$$V_p = \frac{850 \text{ m}^3/\text{s}}{4 \text{ m}\cdot 50 \text{ m}} \cdot 4.25 \text{ m/s}$$

Hence,

$$V_m = \frac{V_p}{V_r} = \frac{4.25}{(80)^{1/2}} = 0.475 \text{ m/s}$$

The model Reynolds number is

$$N_R = \frac{V_m \bar{Y}_m}{\nu} = \frac{0.475\cdot 0.05}{1.1\cdot 10^{-6}} = 21,598$$

which is much greater than the critical Reynolds number (2,000). Hence, the flow is turbulent in the model.

PROBLEMS

8.7.1. A new laboratory site is available for modeling the channel of Example 8.8 so that the length is no longer a restriction, but the roughness coefficient of the material to be used in the movable bed is $n_m = 0.018$. Determine a convenient scale and the corresponding model velocity.

8.7.2. A ship channel model is built to study sedimentation control in the prototype which is 40 m wide and 7.5 m deep and carries a discharge of 300 m³/s. For a vertical scale of 1 : 65 and a roughness coefficient of $n_p = 0.03$, determine all the other needed ratios for the study if $n_m = 0.02$.

8.7.3. A 1 : 100 scale model is constructed to study the pattern of flow in a river reach. If the reach has a Manning's coefficient $n = 0.025$, what should be the corresponding value of n in the model? Discuss the errors that may result from the study if the value of n were not modeled.

8.7.4. A 1 : 25 model is built to study a stilling basin at the outlet of a steep spillway chute. The stilling basin consists of a horizontal floor (apron) with U.S.B.R. Type II baffles installed to stabilize the location of the hydraulic jump. The prototype has a rectangular cross section 25 m wide designed to carry a 75-m³/s discharge. The velocity immediately before the jump is 10 m/s. Determine the following:

(a) The model discharge.

(b) The depth downstream of the jump in the prototype if the dynamic force measured on the model baffles is 16.2 N.

(c) The force on the baffles per unit width of the prototype channel.

(d) The energy dissipated in the basin.

REVIEWING PROBLEMS

8.1 An overflow spillway is designed to be 100 m high and 120 m long, carrying a discharge of 1200 m³/s under an approaching head of 2.75 m. The spillway operation is to be analyzed by an 1:50 model in a hydraulic laboratory. Determine:

(a) The model discharge;

(b) If the discharge coefficient at the model crest measures 2.12, what is the prototype crest discharge coefficient?

(c) If the velocity at the outlet of the model spillway measures 25 m/s, what is the prototype velocity?

(d) If the U.S.B.R. Type II stilling basin, 50 m width is used to dissipate the energy at the toe of the spillway, what is the energy dissipation in the model and in the prototype as measured in units of kilowatts?

(e) What is the efficiency of the dissipator in the prototype?

8.2 An underwater structure is built on the ocean floor where a strong current of 5 m/s is measured. The structure is to be studied by a 1:25 model in a water tunnel using sea water at same temperature as that measured in the ocean. What speed must the water tunnel provide in order to study the force load on the structure due to the current? If the required tunnel velocity is judged to be impractical, can the study be performed in a wide tunnel using air at 20°C? What would be the corresponding air speed required in the tunnel?

8.3 A model is built to study the surface tension phenomenon in a reservoir. Determine the conversion ratios between the model and the prototype for the following quantities if the model is built with a 1:100 scale: (a) rate of flow, (b) energy, (c) pressure, (d) power.

8.4 A concrete caisson 60 m wide, 120 m long, and 12 m high is to be towed in sea water in the longitudinal direction to an off-shore construction site where it will be sunk. The calculated floating depth of the caisson is 8 m, with 4 m remaining above the water surface. A 1:100 model is built to study the operation of the prototype. If the model is towed in a wave tank using sea water, what is the model speed that corresponds to the prototype speed of 1.5 m/s? The model study considers both the skin drag and form drag force (Reynolds number) and the resistance due to the generation of gravity waves in motion (Froude number).

9

WATER MEASUREMENTS

Measurement of water flow provides fundamental data for analysis, design, and operation for every hydraulic system. A variety of devices and methods are available for measuring water flow in the field and in the laboratory. Hydraulic measurements are usually performed to measure velocity, pressure, or discharge and are based on the fundamental laws of physics and of fluid mechanics.

9.1 PITOT TUBES AND CURRENT METERS—VELOCITY MEASUREMENTS

Water flow velocity in every conduit varies between near-zero values close to the stationary wall to a maximum value in the midstream. It is frequently of interest to measure the velocity distribution in a conduit. This is done by making local measurements at several positions in a cross section. The measurements should be made only with velocity probes of small size, so that local flow patterns will not be disturbed by the presence of the probe in the field. Instruments commonly used for velocity measurement are *Pitot tubes* and *current meters*.

Pitot tubes* are tubes bent to measure pressure in the flowing stream. The

---

*Henri de Pitot (1695–1771) first used an open-ended glass tube with a 90°-bend to measure velocity distribution in the River Seine. The rise of water elevation in the vertical part of the tube indicated static pressure. Pitot did not apply Bernoulli's principle to obtain the correct velocity as discussed in this section..

probe usually consists of two tubes that are bent in such a way that the open end of one tube is directly opposite to the velocity vector and the other is parallel to the flow. The probe body is aligned in the direction parallel to the flow (Figure 9.1). In order to reduce the size of the probe, the two tubes are usually combined into one concentric construction, that is, a smaller tube is inside a larger one [Figure 9.1(b)].

At the end opening, 0, the flow reaches a stagnant point where the velocity is zero. The pressure sensed at this opening is the stagnant point pressure, or stagnation pressure. At the side openings, 1, the flow velocity, v, is practically undisturbed. These openings sense the dynamic (or ambient) pressure at the site.

Applying Bernoulli's equation between the two positions, 0 and 1, and neglecting the small vertical distance in between, we may write

$$\frac{p_0}{\gamma} + 0 = \frac{p_1}{\gamma} + \frac{V^2}{2g}$$

Hence, the flow velocity can be determined:

$$V^2 = 2g\left(\frac{p_0 - p_1}{\gamma}\right) = 2g\left(\frac{\Delta p}{\gamma}\right) = 2g(\Delta h)$$

or

$$V = \sqrt{2g\,\Delta h} \tag{9.1}$$

Figure 9.1. Schematic representation of a Pitot tube. (a) Separated tubes. (b) Combined tubes.

Notice that the quantity (Δh) indicates the height of the water column shown in the manometer. A different, immersible indicator liquid, such as mercury, is used for measuring the velocity of water. In such a case, the height of the indicator liquid column shown in the manometer must be converted into equivalent height of water column for computing the velocity using Eq. (9.1).

Pitot tubes are widely used to measure the pressure and the velocity of water flow. They are both reliable and accurate, for they involve a simple physical principle and a simple setup. The Pitot tube is especially good for measuring water velocities under conditions in which the exact direction of the stream flow cannot be determined and in which misalignments of the probe to the stream are likely to occur. The Prandtl-type Pitot tube, as shown in Figure 9.1(b), has approximately a 1 % error at an angle of 20° to the direction of the stream flow.

The outer diameter of a Pitot tube is, typically, say 5 mm. The two pressure tubes inside are much smaller. Because of the small diameters of these tubes, caution must be exercised to free them from air bubble entrapment. Surface tension at the interface can produce a predominant effect in the small tubes that may totally inactivate the system.

Example 9.1

A Pitot tube is used to measure velocity at a certain point in a water pipe. The manometer indicates a column of 14.6 cm height. The indicator fluid has specific weight of 1.95. Compute the velocity.

Solution

The column of 14.6 cm indicator fluid (sp. gr. 1.95) is equivalent to a water column of h.

$$h = 14.6 \cdot 1.95 = 28.47 \text{ cm} = 0.2847 \text{ m}$$

Applying Eq. (9.1), we have

$$V = \sqrt{2gh} = \sqrt{2 \cdot 9.81 \cdot 0.2847} = 2.36 \text{ m/s}$$

Current meters are frequently used to measure the speed of water current. There are two different types of current meter: the cup-type and the propeller-type.

The *cup-type current meter* usually consists of from four to six evenly shaped cups mounted radially about a vertical axis of rotation [Figure 9.2(a)]. The current rotates the cups about the axis at a rate proportional to the current speed. Each revolution is registered by a conductor which trans-

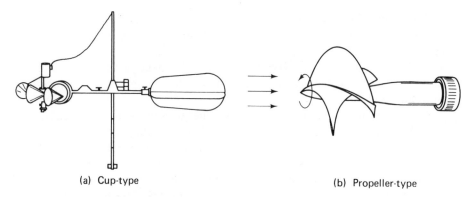

Figure 9.2. Current meters. (a) Cup-type. (b) Propeller-type.

mit electrical contact made by the cup wheel to an earphone or a counter. Most cup-type current meters do not register speed below a few centimeters per second because of starting friction.

The *propeller current meter* has a horizontal axis of rotation [Figure 9.2(b)]. It is more suitable for measuring higher ranges in the midstream regions.

Depending on the design and construction of the current meter, the speed of axial rotation may not be linearly proportional to the speed of the water current. For this reason, each current meter must be individually calibrated before it is used for field measurements. Calibrations may be carried in a straight stretch of a towing tank by towing the probe through still water at constant speeds. A calibration curve covering the range of applicable speeds should be provided with each current meter.

PROBLEMS

9.1.1. A Pitot tube directed into a 4-m/s stream of water has a difference of 14.2 cm in the column in a manometer. Determine the specific gravity of the indicator fluid.

9.1.2. A static tube (with the open end directly opposite to the velocity vector) is inserted into a small stream of flowing oil. Specific gravity of the oil is 0.85. There is a rise of the fluid, h, in the vertical tube. What is the velocity?

9.1.3. A Pitot tube (with both stagnant point pressure and dynamic pressure measured) is used in the stream in Prob. 9.1.2. What is the equivalent height of the oil column?

9.1.4. Measurements in a water tunnel indicate that the stagnation pressure and the dynamic pressure differ by 12.2 cm of mercury column. What is the water velocity?

9.2 PRESSURE MEASUREMENTS

Pressure at any point in a liquid is defi d by the normal force exerted by the liquid on a unit area surface at the point. A common method of measuring pressure is made through openings in the boundary. The liquid is connected through the openings to a bank of manometers in which the value of liquid pressure is determined by the height of liquid in a column. When the liquid is stationary, the manometer measures the hydrostatic pressure at the opening. If the liquid is moving past the boundary, the pressure at the opening decreases with the increase of flow velocity at the opening. The amount of pressure decrease can be calculated by Bernoulli's principle.

It is of paramount importance that the openings on the boundary satisfy certain characteristics, so that the true ambient pressure can be registered. The opening must be flush with the surface and normal to the boundary. Figure 9.3(a) and (b) demonstrates schematically the various correct and incorrect ways of boundary openings for pressure measurement. The plus signs ($+$) indicate that the opening registers a higher than actual pressure value; the minus signs ($-$) indicate that the opening registers a lower than actual pressure value. In order to eliminate the irregularities and variations that might cause significant errors in measurement, a series of piezometric connections may sometimes be made at a given cross section in a closed conduit. The connections are in turn connected into a single manometer column, so that the manometer registers an average pressure of all the piezometric openings in the cross section. This multi-opening system is effective in sections of relatively straight pipe where the velocity profiles are reasonably

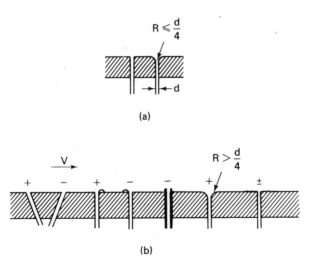

(a)

(b)

Figure 9.3. Pressure openings. (a) Correct connections. (b) Incorrect connections.

symmetric and the pressure difference existing between one side of the pipe and the other is very small. If a large pressure difference exists between any of the openings, measurement error can develop because water may flow out of the higher-pressure side and into the lower-pressure openings. Caution must be taken to ensure that no net flow occurs through any piezometric openings.

$\Delta h = \Delta l \, (\sin \theta)$

Figure 9.4. Inclined manometer.

To increase the sensitivity, inclined manometers may be used in which a small change in pressure can drive the indicator fluid a great distance along the sloped manometer tube (Figure 9.4). Other manometer configurations were discussed in Chapter 1.

The manometer systems may be used to measure pressure in both gases and liquid under steady flow conditions. They are not suitable for applications in time-varying flow fields that require high-frequency response both in probes and in recording systems. Electronic pressure cells are available commercially. The quartz crystal pressure cell, for example, can respond to flow field fluctuation from direct current up to 20,000 Hz over a wide range of pressure. Similar to the piezometric openings, the pressure cells must also be mounted flush with the boundary surface. Slight irregularity in alignment may cause significant error in measurement, as discussed previously.

Example 9.2

An open-ended U-tube (Figure 9.5) is filled with water. Oil is poured into one leg of the U-tube and causes the water surface in one leg to rise 6 cm above the other. The oil column measures 8.2 cm. What is the specific gravity of the oil?

Figure 9.5

Solution

To balance the columns in the two legs of the U-tube, the pressure at levels 1 and 2 must be the same. That is, the pressure generated by the 6-cm water column must be the same as that by the 8.2-cm oil column.

$$p_{water} = 6 \text{ cm}(\gamma_{water}); \qquad p_{oil} = 8.2 \text{ cm}(\gamma_{oil})$$

$$6(\gamma_{water}) = 8.2(\gamma_{oil})$$

$$\text{specific gravity of oil} = \frac{\gamma_{oil}}{\gamma_{water}} = \frac{6}{8.2} = 0.732$$

PROBLEMS

9.2.1. If the U-tube in Example 9.2 is connected to a pressure vessel at leg B and if the column in leg A is mercury instead of oil, what is the pressure in the vessel? End A is open to atmosphere and the vessel contains pressurized air.

9.2.2. A straight circular pipe of 1.2 m diameter has at one section two pressure taps bored into the the vessel wall with exactly the same type of connections. Open-ended mercury U-tubes are connected to the taps to read pressure. If the tap on top of the pipe reads 7.24 cm of mercury column and if that on the bottom reads 16.06 cm, what is the specific gravity of the fluid inside the pipe? What is the mean pressure in the pipe?

9.2.3. A simple, open-ended glass tube is bent 90°. The tube is inserted into a flowing stream with one end facing the direction of the flow and the other end oriented vertically above the water. It is noticed that the water surface in the vertical tube rises 4 cm above the stream surface. What is the stream velocity?

9.3 DISCHARGE MEASUREMENTS IN PIPE FLOWS— VENTURI METERS, NOZZLE METERS, ORIFICE METERS, AND BEND METERS

Although measurement of discharge in pipe flow can be accomplished by several different methods, the simplest and most reliable method is volumetric (or weight) determination. This method involves only a stopwatch and an open tank to collect the water flowing from the pipe. The discharge rate can be determined by measuring the water volume (or weight) collected per unit time. Because of its absolute reliability, this method is frequently used for calibration of various types of flow meter but it cannot be used in routine, operational applications, since the flowing water must be totally diverted into a container tank every time a measurement is made.

The variation in pressure head, associated with localized changes in velocity and energy head loss through a sudden change in pipe cross-sectional geometry, may be correlated to the pipe flow discharge. This principle is used, most commonly, with Venturi meters, nozzle meters, and orifice meters.

A *Venturi meter* is a machine-cast section of pipe having a narrow throat. Two piezometric openings are installed at the entrance and at the throat, as shown in Figure 9.6. Applying the Bernoulli equation at sections 1 and 2, and neglecting head loss, we get

$$\frac{V_1^2}{2g} + \frac{p_1}{\gamma} + z_1 = \frac{V_2^2}{2g} + \frac{p_2}{\gamma} + z_2 \tag{9.2}$$

The continuity equation between the two sections is

$$A_1 V_1 = A_2 V_2 \tag{9.3}$$

where A_1 and A_2 are, respectively, the pipe and the throat cross-sectional areas. Substituting Eq. (9.3) into Eq. (9.2) and rearranging, we have the discharge, Q:

$$Q = \frac{A_1}{\sqrt{\left(\frac{A_1}{A_2}\right)^2 - 1}} \sqrt{2g\left(\frac{p_1 - p_2}{\gamma} + z_1 - z_2\right)} \tag{9.4}$$

The equation may be simplified to

$$Q = C_d A_1 \sqrt{2g\left[\Delta\left(\frac{p}{\gamma} + z\right)\right]} \tag{9.5}$$

where

$$C_d = \frac{1}{\sqrt{\left(\frac{A_1}{A_2}\right)^2 - 1}} \tag{9.5a}$$

For Venturi meters installed in a horizontal position,

$$Q = C_d A_1 \sqrt{2g\left(\frac{\Delta p}{\gamma}\right)} \tag{9.5b}$$

Figure 9.6. Venturi meter.

For Venturi meters, the coefficient, C_d, can be directly computed from the values of A_1 and A_2. The difference between this theoretically computed value and that obtained from experiments should not exceed a few percent under most practical conditions.

For satisfactory operation, the meter should be installed in a section of the pipe where the flow is relatively undisturbed before it enters the meter. To ensure this, a section of straight and uniform pipe, free from fittings, of at least 30 diameters in length must be provided before the meter installation.

Example 9.3

A 6-cm (throat) Venturi meter is installed in a 12-cm diameter horizontal water pipe. The differential manometer shows a reading of 15.2 cm of mercury column. Calculate the discharge.

Solution

$$A_1 = \frac{\pi}{4} \cdot (12)^2 = 113.1 \text{ cm}^2 \qquad A_2 = \frac{\pi}{4} \cdot (6)^2$$

The coefficient, C_d, can be calculated from the area ratio by using Eq. (9.5a).

$$C_d = \frac{1}{\sqrt{\left(\dfrac{A_1}{A_2}\right)^2 - 1}} = 0.577$$

Applying Eq. (9.5b) for a horizontally installed Venturi meter, the discharge, Q, can be calculated.

$$Q = 0.577 \cdot \frac{113.1}{10{,}000}\sqrt{2 \cdot 9.81\left(\frac{15.2 \cdot 13.6}{100}\right)} = 0.0416 \text{ m}^3/\text{s}$$

Nozzle meters and *orifice meters* (Figure 9.7) are based on the same principle of energy head conversion and head loss across a concentric obstruction installed in the pipe conduit. The discharge equations for the nozzle meter and the orifice meter have the same form as that derived for the Venturi meter [Eq. (9.5)]. The main difference in application is that the value of the coefficient of discharge for the nozzle meters and orifice meters would be different from the theoretical value, C_d, calculated by using Eq. (9.5a). This is due primarily to separation of the stream flow from the pipe wall boundary immediately downstream from the obstruction (*vena contrata*).

Nozzle meters and orifice meters especially encounter a significant amount of head loss, as most of the pressure energy that converted to kinetic energy (to rush the fluid through the narrow opening) cannot be recovered. The coefficient of discharge may vary significantly from one meter to another. The value depends not only on the status of flow in the pipe (the pipe Reynolds

Figure 9.7. (a) Nozzle meter. (b) Orifice meter.

number) but also on the area ratio between the nozzle (or orifice) and the pipe, the location of the pressure taps, and the upstream and downstream condition of the pipe flow. For this reason, on-site calibration is recommended for each meter installed. If installation of a pipe orifice meter is made without an on-site calibration, reference should be made to the manufacturer's data, and the detailed installation requirements should be followed.

The coefficient for nozzle meters and orifice meters cannot be directly computed from the area ratio, A_1/A_2. The discharge equations [Eqs. (9.5), (9.5a), and (9.5b)] must be modified by an experimental coefficient, C_v.

$$Q = C_v C_d A_1 \sqrt{2g\, \Delta\!\left(\frac{P}{\gamma} + z\right)} \tag{9.6}$$

where z is the difference in elevations between the two pressure taps. For horizontal installations,

$$Q = C_v C_d A_1 \sqrt{2g\, \Delta\!\left(\frac{P}{\gamma}\right)} \tag{9.6a}$$

Extensive research on the fluid nozzle has been sponsored by the American Society of Mechanical Engineers (ASME) and the International Standards Association (ISA) to standardize the nozzle geometry, installation, specification, and experimental coefficients. One of the typical ASME flow nozzle installations commonly used in the United States and the corresponding experimental coefficients are shown in Figure 9.8.

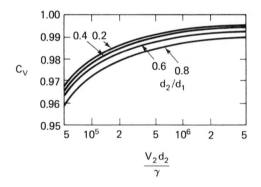

Figure 9.8. ASME nozzle dimensions and coefficients.

Compared to Venturi meters and nozzle meters, orifice meters are even more affected by flow conditions. For this reason, detailed instructions of installation and calibration curves must be provided by the manufacturer for each type and size. Meters installed not strictly according to the instructions should be individually calibrated on site.

Example 9.4

An ASME flow nozzle of 6-cm diameter is installed in a 12-cm waterline. The attached differential manometer reads a 15.2-cm column of mercury. The water in the pipeline is 20°C. Determine the discharge.

Solution

The diameter ratio is $d_2/d_1 = 6/12 = 0.5$, and $C_d = 0.577$ (from Example 9.3). Assume the experimental coefficient $C_v = 0.99$. The corresponding discharge can be calculated from Eq. (9.6a):

$$Q = 0.99 \cdot 0.577 \cdot 0.01131 \sqrt{2 \cdot 9.81 \left(\frac{15.2 \cdot 13.6}{100}\right)} = 0.0411 \ \text{m}^3/\text{s}$$

This value must be verified by checking the corresponding Reynolds number of the nozzle. The N_{R_e} value calculated based on the discharge is

$$N_{R_e} = \frac{\dfrac{0.0411}{\dfrac{\pi}{4}(0.06)^2} \cdot 0.06}{1.007 \cdot 10^{-6}} = 5.2 \cdot 10^4$$

With this Reynolds number value, the chart in Figure 9.8 gives a better value of the experimental coefficient, $C_v = 0.96$. Hence, the correct discharge is

$$Q = \frac{0.96}{0.99} \cdot 0.0411 = 0.0399 \ \text{m}^3/\text{s}$$

A *bend meter* (Figure 9.9) measures the pressure difference between the outer and inner sides of a bend in the pipeline. Centrifugal force developed at a pipe bend forces the mainstream to flow closer to the outer wall of the pipe at the bend. Difference in pressure is developed between the inside and outside of the bend. The pressure difference increases as the flow rate increases. The relationship between the pressure difference measured and the discharge in the pipe can be calibrated for flow rate determinations. The discharge equation may be expressed as

$$Q = C_d A \sqrt{2g \left[\Delta\left(\frac{P_o}{\gamma} - \frac{P_i}{\gamma}\right)\right]} \tag{9.7}$$

where A is the pipe cross-sectional area and P_i and P_o are the local pressure

Figure 9.9. Bend meter.

values registered at the inside and outside of the pipe bend, respectively. C_d is the discharge coefficient, which can be determined by calibration in place.

If the bend meter cannot be calibrated in place, the pipe discharge may still be determined with an accuracy of approximately 10% if the pipe flow Reynolds number is sufficiently large and if at least 30 diameters of straight pipe are provided upstream from the bend. In such case, the discharge coefficient is approximated by

$$C_d = \frac{R}{2D} \qquad (9.8)$$

where R is the centerline radius at the bend and D is the pipe diameter.

Bend meters are inexpensive. An elbow already in the pipeline may be used without additional installation cost or causing added energy head loss.

PROBLEMS

9.3.1. The maximum flow rate in a 20-cm diameter waterline is 0.12 m³/s. If a 12-cm (throat) Venturi meter is installed in the pipe to measure the flow, estimate how long a vertical U-tube must be provided for the differential manometer if water and mercury are used in the manometer.

9.3.2. A 20-cm Venturi meter is installed in a 50-cm diameter waterline. If the pressure taps read 30 N/cm² and 20 N/cm² when the meter is installed in a horizontal position, what is the flow rate?

9.3.3. A 10-cm ASME flow nozzle is installed in a 20-cm waterline. The attached manometer contains mercury and water and registers a difference of 42 cm column. Calculate the discharge of the pipe.

9.3.4. Determine the discharge in the 40-cm diameter waterline shown in Figure 9.10. The nozzle meter has a throat diameter of 32 cm and is built and installed according to the ASME standard.

Figure 9.10

9.3.5. A 30-cm orifice plate is installed in a 50-cm diameter, horizontal waterline. In a laboratory calibration, 6.78 m³ of water were collected in 7.0 s. The water mercury manometer read a difference of 18.24 cm in a vertical mercury column. What is the discharge coefficient of the orifice meter?

9.3.6. A 5 cm orifice meter is installed in a horizontal 10 cm dimeter pipeline. A differential mercury manometer shows a deflection of 9.0 cm which include the head loss. Assuming that the pressure energy converted to kinetic energy at the orifice (to speed up the water flow) is totally lost through the meter, determine the discharge in the pipe. If a carefully streamlined venturi meter is used instead of the orifice meter, and it allows 60% of the pressure energy to recover, what would be the deflection of mercury column for the same discharge?

9.3.7. A bend meter installed in a 75-cm diameter water pipe is shown in Figure 9.9. The installation delivered 48 m³ of water in 1 min. Determine the deflection in cm of mercury when the meter is in a horizontal position.

9.3.8. Determine the coefficient of discharge for the bend meter in Prob. 9.3.7 if the meter is installed in a vertical position for the same discharge. Note that in the vertical position the pressure taps are mounted along a 45° line from the horizontal datum.

9.4 DISCHARGE MEASUREMENTS IN OPEN CHANNELS—WEIRS, GATES, AND PARSHALL FLUMES

Essentially, a weir is a simple overflow structure extending across a channel and normal to the direction of the flow. Various types of weirs exist, and they are generally classified by shape. Weirs may be either sharp-crest or broad-crest.

Sharp-Crest Weirs Sharp-crest weirs include four basic types: (1) the horizontal weir without end contractions, (2) the horizontal weir with end contractions, (3) the V-notch weir, and (4) the trapezoidal weir (see Figure 9.11).

The *uncontracted horizontal* weir extends across the entire width of a uniform reach in the channel. A standard uncontracted, horizontal weir should meet the following requirements:

1. The crest of the weir should be horizontal, sharp-edged, and normal to the flow.
2. The weir plate should be vertical and have a smooth upstream surface.
3. The approach channel should be uniform and the water surface should be free from large surface waves.

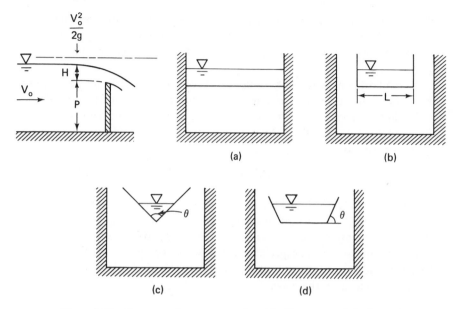

Figure 9.11. Common sharp-crest weirs. (a) Uncontracted horizontal weir. (b) Contracted horizontal weir. (c) V-notched weir. (d) Trapezoidal weir.

The basic discharge equation for a standard uncontracted, horizontal weir is:

$$Q = C_d L H^{3/2} \qquad (9.9)$$

where L is the length of the crest, H is the head on the weir, and C_d is the discharge coefficient, which is derived from experimental data by the U.S. Bureau of Reclamation. Using the British units (H, L, p in feet, and Q in cubic feet per second), the discharge coefficient may generally be expressed as:

$$C_d = 3.22 + 0.40\frac{H}{p} \qquad (9.10)$$

where p is the weir height.

The *contracted horizontal weir* has a crest that is shorter than the width of the channel, so that water must be contracted both horizontally and vertically in order to flow over the crest. The weir may be contracted at either one end or at both ends. The general discharge equation may be expressed as

$$Q = C_d\left(L - \frac{nH}{10}\right)H^{3/2} \qquad (9.11)$$

where n is the number of contractions at the end [$n = 1$ for contraction at one end and $n = 2$ for contraction at both ends, as shown in Figure 9.11(b)]. The coefficient of discharge, C_d, can only be effectively determined by calibration in place.

A standard contracted horizontal weir is one whose crest and sides are so far removed from the bottom and sides of the channel that full contraction is developed. The dimension of a standard contracted horizontal weir is shown in Figure 9.12. The discharge equation of the standard weir is given by the USBR* as

$$Q = 3.33(L - 0.2H)H^{3/2} \qquad (9.12)$$

This formula is developed for the British measurement system, with L and H in feet and Q in the unit of cubic feet per second. The formula can be derived for the use of the S. I. units, as demonstrated in Example 9.5.

The triangular weir is especially useful when accuracy in measurement is desired for a large range of water elevations. The discharge equation for a triangular weir takes the general form of

$$Q = C_d\left(\tan\frac{\theta}{2}\right)H^{5/2} \qquad (9.13)$$

where θ is the weir angle, as shown in Figure 9.11(c), and the discharge coefficient, C_d, is determined by calibration in place.

A standard USBR 90°-*V-notch weir* consists of a thin plate. The sides of the notch are inclined 45° from the vertical. The weir operates as a contracted horizontal weir and all conditions for accuracy stated for the standard contracted horizontal weir apply. The minimum distances of the sides of the weir from the channel banks should be at least twice the head on the weir, and the

Figure 9.12. USBR standard weirs. (a) Contracted horizontal weir. (b) 90°-V-notch weir.

*Water Measurement Manual, U.S. Bureau of Reclamation, 1967.

minimum distance from the crest to the channel bottom should be at least
twice the head on the weir. The discharge equation of the standard 90°-V-
notch weir is given by the USBR as

$$Q = 2.49H^{2.48} \qquad (9.14)$$

This formula, like Eq. (9.12), is developed for the British measurement
system. Units used in the formula are restricted to feet and to cubic feet per
second.

The *trapezoidal weir* has hydraulic characteristics in between that of the
contracted horizontal weir and the triangular weir. The general discharge
equation developed for the contracted horizontal weir may be applied to the
trapezoidal weir with an individually calibrated discharge coefficient.

A USBR standard trapezoidal weir is also known as the *Cipolletti weir*. It
has a horizontal crest and the sides incline outwardly at a slope of 1 (horizontal) to 4 (vertical). All conditions for accuracy stated for the standard
weirs described above apply to the Cipolletti weir. The height of the weir
crest should be at least twice the head of the approach flow above the bottom,
and the distances from the sides of the notch to the sides of the channel
should also be at least twice the head. The discharge equation for the Cipolletti weir is given by the USBR as

$$Q = 3.367LH^{3/2} \qquad (9.15)$$

As in Eqs. (9.12) and (9.14), the units for the above formula are restricted to
feet and cubic feet per second. The standardized dimension of the Cipolletti
weir is given in Figure 9.13.

Figure 9.13. USBR standard trapezoidal weir.

Example 9.5

Laboratory measurement made on a USBR standard contracted horizontal
weir shows a discharge of 0.25 m³/s under a head of $H = 0.2$ m. Determine
the discharge coefficient in the given units.

Solution

Applying Eq. (9.11) for the contracted horizontal weir, we get

$$Q = C_d\left(L - \frac{nH}{10}\right)H^{3/2}$$

Here, $L = 1.56$ m, $H = 0.2$ m, and $n = 2$ for contraction at both ends;

$$0.25 = C_d\left(1.56 - \frac{2 \cdot 0.2}{10}\right)(0.2)^{3/2}$$

$$C_d = 1.84$$

Broad-Crest Weirs   A broad-crest weir provides a stretch of elevated channel floor over which the critical flow takes place (Figure 9.14). Depending on the height of the weir in relation to the depth of the approaching channel, the discharge equation may be derived from the balance of forces and momentum between the upstream approach section 1 and the section of minimum depth 2 on the crest of the weir. For unit width of the weir, the following equation may be written:

$$\rho q\left(\frac{q}{d_2} - \frac{q}{d_1}\right) = \frac{1}{2}\gamma d_1^2 - \frac{1}{2}\gamma d_2^2 - \frac{1}{2}\gamma h(2d_1 - h) \qquad (9.16)$$

The conditions provided above are not sufficient for us to simplify Eq. (9.16) into a one-to-one relationship between the approach water depth and the discharge. An additional equation was obtained from experimental measurements* for the average flow:

$$d_1 - h = 2d_2 \qquad (9.17)$$

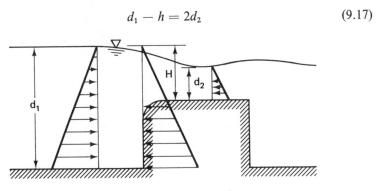

Figure 9.14. Broad-crest weir.

*H. A. Doeringsfeld and C. L. Barker, Pressure-momentum theory applied to the broad-crested weir, Trans. ASCE, vol. 106, pp. 934–946, 1941.

Substituting Eq. (9.17) into Eq. (9.16) and simplifying, we have

$$q = 0.433\sqrt{2g}\left(\frac{d_1}{d_1 + h}\right)^{1/2}H^{3/2} \tag{9.18}$$

The total discharge over the weir is

$$Q = Lq = 0.433\sqrt{2g}\left(\frac{d_1}{d_1 + h}\right)^{1/2}LH^{3/2} \tag{9.19}$$

where L is the length of the weir crest and H is the height of the approach water above the crest.

Considering the limit of a weir with zero height ($h = 0$) to infinity ($h \rightarrow \infty$), Eq. (9.19) may vary from $Q = 1.92LH^{3/2}$ to $Q = 1.36LH^{3/2}$ with the units in meters and seconds.

Venturi Flumes   The use of a weir is probably the simplest method for measuring discharge in open channel flows. The disadvantage, however, is the relatively high energy loss and the sedimentation deposited in the pool immediately upstream of the weir. These difficulties can be partially overcome by using a *critical flow flume* (*Venturi flume*).

A variety of Venturi flumes have been designed for field application. Most of the flumes operate with a submerged outflow condition and have a critical depth created at a contracted section (throat) and a hydraulic jump at the exit. The discharge through the flume can be calculated by reading the water depth from the observation wells located at the critical flow section and another reference.

The most extensively used critical flow flume in the United States is the *Parshall flume*, developed by R. L. Parshall* in 1920. The flume was experimentally developed for the British measurement system. It has fixed dimensions as shown in Figure 9.15 and Table 9.1. Empirical discharge equations were developed to correspond to each flume size. These equations are listed below:

Throat Width	Discharge Equation	Free Flow Capacity (cfs)
3 in.	$Q = 0.992H_a^{1.547}$	0.03–1.9
6 in.	$Q = 2.06H_a^{1.58}$	0.05–3.9
9 in.	$Q = 3.07H_a^{1.53}$	0.09–8.9
1 to 8 ft	$Q = 4WH_a^{(1.522W)^{0.026}}$	Up to 140
10 to 50 ft	$Q = (3.6875W + 2.5)H_a^{1.6}$	Up to 2000

*R. L. Parshall and C. Rohwer, *The Venturi Flume*, Colorado Agricultural Experimental Station Bulletin No. 265 (1921). R. L. Parshall, "The Improved Venturi Flume," *Trans. ASCE*, **89** (1926), 841–851.

TABLE 9.1 Parshall Flume Dimensions

W		A		2/3 A		B		C		D		E		F		G		H		K		M		N		P		R		X		Y		Z		FREE-FLOW CAPACITY	
FT	IN	FT	IN	FT	IN	FT	IN	FT	IN	FT	IN	FT	IN	FT	IN	FT	IN	FT	IN	FT	IN	FT	IN	FT	IN	FT	IN	FT	IN	FT	IN	FT	IN	FT	IN	MIN. SEC-FT	MAX. SEC-FT
0	1 [1]	1	2 9/32	0	9 17/32	1	2	0	3 21/32	0	6 19/32	0	6 to 9	0	3	0	8	0	8 1/8	0	3/4	—	—	0	1 1/8	—	—	—	—	0	5/16	0	1/2	0	1/8	0.01	0.19
0	2 [2]	1	4 5/16		10 7/8	1	4	0	5 5/16	0	8 13/32	0	6 to 10	0	4 1/2	0	10	0	10 1/8	0	7/8	—	—	0	1 11/16	—	—	—	—	0	5/8	0	1	0	1/4	0.02	0.47
0	3 [3]	1	6 3/8	1	1/4	1	6	0	7	0	10 3/16	1 to 1 1/2		0	6	1	0	1	5 5/32	0	1	—	—	0	2 1/4	—	—	—	—	0	1	0	1 1/2	0	1/2	0.03	1.13
0	6	2	7/16	1	4 5/16	2	0	1	3 1/2	1	3 5/8	2	0	1	0	2	0	—	—	0	3	1	0	0	4 1/2	2	11 1/2	1	4	0	2	0	3	—	—	0.05	3.9
0	9	2	10 5/8	1	11 1/8	2	10	1	3	1	10 5/8	2	6	1	0	1	6	—	—	0	3	1	0	0	4 1/2	3	6 1/2	1	4	0	2	0	3	—	—	0.09	8.9
1	0	4	6	3	0	4	4 7/8	2	0	2	9 1/4	3	0	2	0	3	0	—	—	0	3	1	3	0	9	4	10 3/4	1	8	0	2	0	3	—	—	0.11	16.1
1	6	4	9	3	2	4	7 7/8	2	6	3	4 3/8	3	0	2	0	3	0	—	—	0	3	1	3	0	9	5	6	1	8	0	2	0	3	—	—	0.15	24.6
2	0	5	0	3	4	4	10 7/8	3	0	3	11 1/2	3	0	2	0	3	0	—	—	0	3	1	3	0	9	6	1	1	8	0	2	0	3	—	—	0.42	33.1
3	0	5	6	3	8	5	4 3/4	4	0	5	1 7/8	3	0	2	0	3	0	—	—	0	3	1	3	0	9	7	3 1/2	1	8	0	2	0	3	—	—	0.61	50.4
4	0	6	0	4	0	5	10 5/8	5	0	6	4 1/4	3	0	2	0	3	0	—	—	0	3	1	6	0	9	8	10 3/4	2	0	0	2	0	3	—	—	1.3	67.9
5	0	6	6	4	4	6	4 1/2	6	0	7	6 5/8	3	0	2	0	3	0	—	—	0	3	1	6	0	9	10	1 1/4	2	0	0	2	0	3	—	—	1.6	85.6
6	0	7	0	4	8	6	10 3/8	7	0	8	9	3	0	2	0	3	0	—	—	0	3	1	6	0	9	11	3 1/2	2	0	0	2	0	3	—	—	2.6	103.5
7	0	7	6	5	0	7	4 1/4	8	0	9	11 3/8	3	0	2	0	3	0	—	—	0	3	1	6	0	9	12	6	2	0	0	2	0	3	—	—	3.0	121.4
8	0	8	0	5	4	7	10 1/8	9	0	11	1 3/4	3	0	2	0	3	0	—	—	0	3	1	6	0	9	13	8 1/4	2	0	0	2	0	3	—	—	3.5	139.5
10	0	—	—	6	0	14	0	12	0	15	7 1/4	4	0	3	0	6	0	—	—	0	6	—	—	1	1 1/2	—	—	—	—	0	9	1	0	—	—	6	200
12	0	—	—	6	8	16	0	14	8	18	4 3/4	5	0	3	0	8	0	—	—	0	6	—	—	1	1 1/2	—	—	—	—	0	9	1	0	—	—	8	350
15	0	—	—	7	8	25	0	18	4	25	0	6	0	4	0	10	0	—	—	0	9	—	—	1	6	—	—	—	—	0	9	1	0	—	—	8	600
20	0	—	—	9	4	25	0	24	0	30	0	7	0	6	0	12	0	—	—	1	0	—	—	2	3	—	—	—	—	0	9	1	0	—	—	10	1000
25	0	—	—	11	0	26	0	29	4	35	0	7	0	6	0	13	0	—	—	1	0	—	—	2	3	—	—	—	—	0	9	1	0	—	—	15	1200
30	0	—	—	12	8	26	0	34	8	40	4 3/4	7	0	6	0	14	0	—	—	1	0	—	—	2	3	—	—	—	—	0	9	1	0	—	—	15	1500
40	0	—	—	16	0	27	0	45	4	50	9 1/4	7	0	6	0	16	0	—	—	1	0	—	—	2	3	—	—	—	—	0	9	1	0	—	—	20	2000
50	0	—	—	19	4	27	0	56	8	60	9 1/4	7	0	6	0	20	0	—	—	1	0	—	—	2	3	—	—	—	—	0	9	1	0	—	—	25	3000

1) Tolerance on throat width (w) ± 1/64 inch; tolerance on other dimensions ± 1/32 inch. Sidewalls of throat must be parallel and vertical

2) From Colorado State University Technical Bulletin No. 61

3) From US Department of Agriculture Soil Conservation Circular No. 843

4) From Colorado State University Bulletin No. 426-A

Plan

Section L-L

Figure 9.15. Parshall flume dimensions. (Courtesy of U.S. Bureau of Reclamation.)

In the above equations Q is the discharge in cubic feet per second (cfs), W is the throat width, and H_a is the water level reading from the observation well, a, measured in feet. Readers should notice that these are derived strictly for the units stated and for the dimensions specified in Figure 9.15 and Table 9.1. These equations do not have an equivalent version for the metric system.

When the ratio of gage reading H_b (from observation well B) to gage reading H_a (from observation well A) exceeds the following values:

> 0.50 for flumes 1, 2, and 3 in. wide
> 0.60 for flumes 6 in. and 9 in. wide
> 0.70 for flumes 1 ft to 8 ft wide
> 0.80 for flumes 8 ft to 50 ft wide

the flow is said to be submerged. The effect of the downstream submergence is to reduce the discharge through the flume. In this case, the discharge computed by the above equations should be corrected by Considering both readings H_a and H_b.

Figure 9.16 shows the rate of submerged flow through a 1 ft Parshall flume. The diagram is made applicable to larger flumes (up to 8 ft) by multiplying the corrected discharge with a factor given below for the particular size selected:

Flume Size (ft)	1.0	2.0	3.0	4.0	6.0	8.0
Correction factor	1.0	1.8	2.4	3.1	4.3	5.4

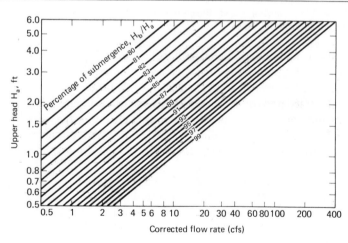

Figure 9.16. Flow rate correction for 1 ft submerged Parshall Flume (after R. L. Parshall, 1950*).

Flume Size (ft)	10	15	20	30	40	50
Correction factor	1.0	1.5	2.0	3.0	4.0	5.0

Figure 9.17. Flow rate correction for 10 ft submerged Parshall Flume (after R. L. Parshall, 1950*)

*R. L. Parshall, *Measuring Water in Irrigation Channels with Parshall Flumes and Small Weirs.* U.S. Soil Conservation Service, Circular 843 (1950). R. L. Parshall, *Parshall Flumes of Large Size.* Colorado Agricultural Experimental Station Bulletin No. 426A (1953).

Figure 9.17 shows the rate of submerged flow through a 10 ft Parshall flume. The diagram is also made applicable to larger size flumes (up to 50 ft) by multiplying the corrected discharge for the 10 ft flume by a factor given below for the particular size used.

PROBLEMS

9.4.1. A contracted horizontal weir with a crest 4 m long has a head 1.1 m above the crest. Determine the discharge if the height of the crest is 3 m.

9.4.2. A rectangular sharp-crest weir with end contractions has a crest 2.0 m long. How high should it be placed in a channel to maintain an upsteam depth of 2.25 m for a discharge of 1 m³/s?

9.4.3. The discharge in a rectangular channel 4 m wide is constant. A depth of 2.3 m is maintained by a 1.7-m high rectangular contracted weir with a 1-m long horizontal crest. This weir is to be replaced by an uncontracted horizontal weir that will maintain the same upstream depth. Determine the height of the weir.

9.4.4. Flow occurs over a sharp-crest, uncontracted horizontal weir 1-m high under a head of 0.3 m. If the weir is replacing a weir just one-half of its height, what change in depth of water can be expected in the upstream channel?

9.4.5. Laboratory tests on a 60°-V-notch weir gave the following results: $H = 0.3$ m, $Q = 0.022$ m³/s, $H = 0.6$ m, and $Q = 0.132$ m³/s. Determine the formula of weir discharge.

9.4.6. A broad-crest rectangular weir is 1 m high and 3 m long. The weir is constructed with a well-rounded upstream corner and a smooth surface. What is the discharge if the head is 0.4 m?

9.4.7. Calculate the flow rate over a USBR standard contracted horizontal weir 3 m long which has a head of 0.6 m.

9.4.8. What length Cipoletti weir (USBR standard trapezoidal weir) is needed for a flow up to 0.8 m³/s if the maximum head is limited to 0.25 cm?

9.4.9. Determine the discharge in cubic meters per second measured by a 10 ft-Parshall flume if the gage reading H_a is 1 m at a free-flow condition.

9.4.10. Determine the discharge through a 4-ft Parshall flume if the gage reading from well a is 1.0 m and from well b is 0.8 m.

9.4.11. A 40-ft Parshall flume measures $H_a = 1.0$ m and $H_b = 0.95$ m; determine the discharge.

10

ELEMENTS IN HYDROLOGY

The science of hydrology deals with the occurrence and distribution of water in the various regions of the earth. It is important for engineers to understand the basic principles of hydrology and the methods of data analysis so that they can make quantitative estimations of water occurrence in their designs of hydraulic structures. Water in the atmosphere continuously undergoes the cycle of evaporation, nucleation, transportation, condensation, and precipitation. Regional and seasonal distributions of water precipitation on land surface, its runoff in a stream, overland flow, or infiltration into the ground aquifer and reappearance in a stream, are topics of concern to hydraulic engineers.

10.1 PRECIPITATION

The earth's atmosphere contains water in the gaseous state as a result of transpiration from plants and evaporation from ocean and land surfaces. The amount of water vapor that any air parcel can hold depends on the temperature. The maximum moisture contained in air at a given temperature is known as the *saturation absolute humidity*. Since the saturation absolute humidity is a function of temperature, water vapor in the air is condensed from the gaseous state when the air is cooled below a certain temperature.

The condensed moisture returns to the earth's surface in liquid state in various forms of precipitation.

Atmospheric temperature decreases with altitude. Cooling of a parcel of air can be effectively achieved by lifting the air to a higher altitude. Lifting of air in the atmosphere can occur through mechanical lifting and thermodynamical lifting.

When winds carry moist air (air parcel with near-saturation moisture content) from a lower elevation to a region where the surface of the earth is at a higher elevation, such as from the ocean surface to a mountain range, the air is forced to gain altitude in order to pass over the highlands. The increase in altitude causes an expansion and lowers the pressure, which, in turn, results in a lower temperature. The lowering of air temperature results in an increase of relative humidity (or decrease in saturation absolute humidity). When the air temperature is lowered to a point where the saturation absolute humidity is reached, water vapor condenses and precipitation takes place. This mechanical lifting process, known as *orographic lifting*, is largely responsible for the high seasonal precipitation on the Pacific Northwest coast of the United States.

Thermodynamical lifting results from the unequal heating of the earth's surface which causes instability in the atmosphere. Convective currents bring the moist air to heights that permit condensation. Precipitation resulting from this process can cause localized, heavy (flood-producing) rainfall. The seasonal, flood-producing rainfall along the upper Gulf Coastal plains is a good example. Figure 10.1 shows the regional and seasonal distributions of precipitation in the continental United States.

Hydraulic engineers often need to estimate rainfall magnitudes for design purposes. The magnitude of maximum rainfall for a particular duration is necessary for the design of flood control or of storm drainage, and the minimum rainfall and the drought period must be analyzed for the design of large-scale water storage or water supply systems.

Measurement of precipitation is made by distributing rain gages on the watershed. If the gages are evenly distributed in the watershed, a single arithmetic mean can suffice for the average depth in estimating the precipitation volume on the watershed (Figure 10.2). If the areal distribution of gages on the watershed is uneven, a method of weighing the precipitation is necessary for obtaining a true average valve. The simplest method of weighing is to use either the *Thiessen method* or the *isohyetal method*.

The Thiessen method provides a weighing factor for each gage in the watershed area. The gage stations are plotted on a map and straight lines are drawn to connect all the stations (Figure 10.3). Perpendicular bisectors of these connection lines form polygons around each station and identify the respective effective area for the station. The area of each polygon is deter-

Figure 10.1. Seasonal and regional distribution of precipitation in the continental United States (in centimeters). (Courtesy of U.S. Environmental Data Service.)

Figure 10.2. Arithmetic method of weighing precipitation.

Figure 10.3. Thiessen method of weighing precipitation.

mined and expressed as a percentage of the total area. Weight average precipitation in the total area is computed by multiplying the precipitation measured at each station by its percentage of area and then totaling. The weighing method usually gives more accurate results than those obtained by simple arithmetical averaging. The Thiessen method assumes linear variation of precipitation between stations and does not allow the consideration of localized orographical influences. This method, therefore, is suitable only for relatively small watershed areas.

The most accurate method for averaging precipitation over an area is the isohyetal method (Figure 10.4). The amount of precipitation measured at each gage station is registered for each of the stations on a suitable map. Contours of equal precipitation (isohyets) are then drawn. The areas between

Figure 10.4. Isohyet method of weighing precipitation.

any two isohyets are determined by planimetry. The average precipitation for an area is computed by weighing the average precipitation between successive isohyets (usually taken as the average value of the two successive isohyets) with the area factor between the isohyets. These products are then totaled.

Example 10.1

Compute the average rainfall intensity over the watershed areas as shown in Figures 10.2, 10.3, and 10.4 by (a) the arithmetic method, (b) the Thiessen method, and (c) the isohyet method.

Solution

(a) Arithmetic method:

$$\frac{133 + 115 + 110}{3} = 119.3 \text{ mm}$$

(b) Thiessen method:

	Gaged Precip	Area Factor	Weighted Precip
I.	110	0.30	33
II.	107	0.07	7.5
III.	115	0.30	34.5
IV.	118	0.09	10.6
V.	133	0.16	21.3
VI.	121	0.08	9.7
		1.00	116.6 mm

(c) Isohyet method:

	Avg. Precip.	Area Factor	Weighed Precip.
1.	109	0.18	19.6
2.	112.5	0.36	40.5
3.	117.5	0.18	21.2
4.	122.5	0.11	13.5
5.	127.5	0.08	10.2
6.	133.0	0.09	12.0
		1.00	117 mm

PROBLEMS

10.1.1. The average annual precipitation for the four subbasins constituting a large drainage basin is 123.7 cm, 114.3 cm, 125.5 cm, and 98.6 cm. The areas covered are, respectively, 520 km², 774 km², 350 km², and 676 km². What is the annual average precipitation for the whole drainage area?

10.1.2. Precipitation Station X was inoperative for repair for a period of one month. A storm occurred during this time. The precipitation recorded during the storm at three surrounding stations, A, B, and C, were, respectively, 15.3 cm, 17.1 cm, and 14.0 cm. The normal annual precipitation amounts at Stations X, A, B, and C are, respectively, 137 cm, 155.7 cm, 182.8 cm, and 136.8 cm. Discuss possible ways for estimating the storm precipitation for station X.

10.1.3. Precipitation recorded at the six stations shown in Figure 10.5 indicates the distribution during a particular storm. Determine the average precipitation for the entire area by (a) the arithmetic method, (b) the Thiessen method, and (c) the isohyetal method.

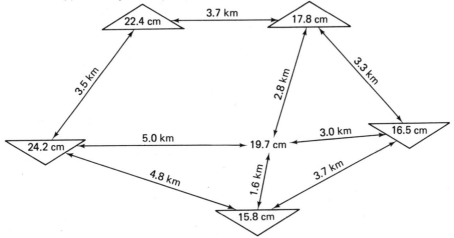

Figure 10.5

10.2 RUNOFF

As precipitation falls toward the earth's surface, a portion of it is intercepted by vegetation (*interception*), surface retention (*depression storage*), and soil moisture. The water intercepted is retained by the leaves and stems of vegetation. Depression storage includes water retained as puddles in surface depressions. Soil moisture is held as capillary water in the smaller pores of the soil or as hydroscopic water absorbed on the surface of soil particles.

The excess water may follow two major paths to a stream. A portion travels as overland flow across the ground surface to the nearest channel and is commonly known as *surface runoff*. Other water may infiltrate the soil and flow under the water table to an open stream channel as effluent flow.

Overland flow is usually considered *direct runoff*. This water reaches the stream channel shortly after it falls to the earth's surface and it is completely discharged from a drainage basin within a few days. The portion of water which percolates downward through the soil and reaches the ground water may eventually reach the open stream channel, is called *ground water runoff*. Usually, it takes a longer period of time for the ground water runoff to reach the stream. The characteristics of direct runoff and groundwater runoff differ so greatly that they should be treated separately in analyzing a particular runoff phenomenon. Unfortunately, there is no practical means of differentiating between ground water runoff and direct runoff after they have reached the stream. The technique of *hydrograph* synthesis was developed as a standard means of runoff analysis.

Basic data required for the construction of a hydrograph usually include a stream stage and a stream discharge as functions of time.

The simplest way to measure a river stage is to use a staff gage, which is a vertical scale fixed to the bottom of the channel so that a portion of the gage is above the water surface at all times [see Figure 10.6(a)]. Continuous recording of the river stage can be made by using a float in connection with chronographic water-stage recorder [Figure 10.6(b)]. As the river stage varies, the motion of the cable is measured and is automatically plotted on a chronographic recording sheet which moves at a constant speed with respect to the recording marker.

Several methods of discharge measurement in open channels have been discussed in Chapter 9. In common practice, however, it is frequently necessary to assess the discharge rate at a stream location where a measuring structure, such as a weir or a Venturi flume, is not available. In this case, the stream discharge can be measured by using the following procedure.

The procedure involves dividing the stream cross section into a number of vertical sections (Figure 10.7). Flow velocity varies approximately from zero at the channel bottom to a maximum value near the surface. Many field test results have shown that the variation is such that the average velocity for each section can be represented by the average of the velocities measured by a

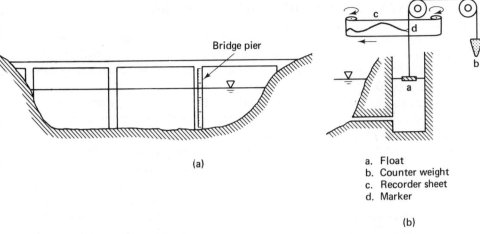

a. Float
b. Counter weight
c. Recorder sheet
d. Marker

(a)

(b)

Figure 10.6. Measurement of river stage. (a) Staff gage. (b) Water-stage recorder.

Figure 10.7. Procedure for discharge measurement with current meter.

current meter located at 0.2 and 0.8 depth below the surface. The average velocity in each vertical section is multiplied by the cross-sectional area to give the discharge in the section. Summation of discharges from all the sections gives the total discharge in the channel.

The accuracy of the measurement increases with the number of vertical sections. The accuracy must be limited, however, to those measurements which can be made within a reasonable time interval. This is especially true if the stage is changing rapidly because the discharge may change with the stage.

PROBLEM

10.2.1 The following data were collected in the river cross section shown in Figure 10.7 which is drawn with a scale of 1 cm = 2.5 m. Determine the discharge of the river. Velocities are expressed in meters per second.

0.2 d		4.0	5.3	6.3	6.5	6.6	6.8	6.5	6.4	5.8	5.4
0.8 d	2.3*	3.4	4.3	5.3	5.8	5.9	5.9	5.6	5.4	4.0	3.2

*One point measurement representing the average velocity.

10.3 HYDROGRAPH AND UNIT HYDROGRAPH

A *hydrograph* is a graphical presentation of the time distribution of discharge at a particular location in the flow path (the stream). A typical hydrograph resulting from a single storm consists of a rising limb, a peak, and a recession (see Figure 10.8).

In analyzing and designing hydraulic structures it is frequently of engineering interest to separate the direct surface runoff from the ground water contributions in a hydrograph. One of the simplest methods, and one that is as easily justified as any other, is to construct a straight line from the point at which the rising limb begins to the point of the greatest curvature on the recession side. The area between the curve and the straight line represents the volume of direct surface runoff. The area below the straight line represents the contribution to the stream flow from ground water runoff, also known as the *base flow*. An example of the straight-line hydrograph separation is shown in Figure 10.8.

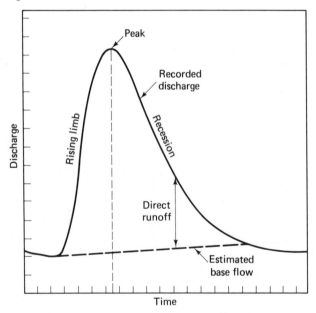

Figure 10.8. Typical hydrograph (separation of base flow).

The time-dependent surface runoff may be expressed in terms of an equivalent depth resulting from a uniformly distributed precipitation over the watershed area. Such considerations lead to the concept of the unit hydrograph.

The basic principles of the unit storm hydrograph were introduced by

Sherman* and are illustrated in Figure 10.9(a). The method considers a net rainfall that occurs at a uniform rate during a specific period of time. The time subdivision is arbitrarily defined according to the rainfall effective duration (t_r). If the runoff volume resulting from this rainfall is equivalent to a depth of 1 cm of water over the whole basin, then the hydrograph from this storm is defined as a *unit hydrograph*. A storm that has exactly such a specific value of runoff is not likely to occur, but the concept of the unit hydrograph has been used in practice as a common tool to determine the actual stream flow. Example 10.2 shows the general procedure for determining a unit hydrograph from a nearly uniform storm.

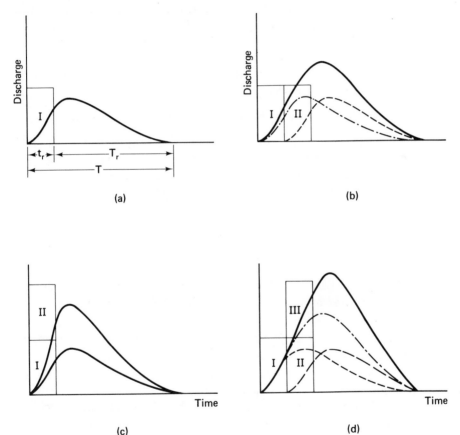

Figure 10.9. Composite hydrograph. (a) Unit storm. (b) Two successive unit storms. (c) Two-unit storm. (d) One-unit storm followed by two-unit storm.

*L. K. Sherman, Streamflow from Rainfall by Unit-Graph Method, *Engr. News Record*, *108* (1932), 501–505.

A unit hydrograph is constructed for a particular storm and defined for a particular drainage area designated. The unit hydrograph will be different for other unit storms and other drainage areas. The time duration of a unit hydrograph, i.e., the time period required to drain the storm from the area, is called the base width (T). This includes the period of the storm (t_r) and the time duration of discharge after the end of the storm, which is known as the recession time (T_r). The recession time is equal to the time required for the water from the most distant point in the drainage area to reach the location of the hydrograph.

Normally, a storm may occur in any distribution of intensity and duration. A composite hydrograph can be constructed by approximation of the actual storm pattern with a variety of ways of combining more than one unit hydrographs. A few typical examples of composite hydrographs are shown in Figure 10.9(b), (c), and (d).

The basic relationship between a unit hydrograph and an actual hydrograph in a particular drainage area may thus be expressed as follows:

$$\text{net runoff} \times \text{unit hydrograph} = \text{direct surface discharge}$$

In applying an existing unit hydrograph, we need only measure the net runoff (cubic meters) to determine the discharge. In deriving the unit hydrograph, however, we must carefully measure both the net runoff and the direct runoff hydrograph to determine the ordinates of a unit hydrograph.

The derivation of a unit hydrograph involves basically four steps:

1. Selection of a reasonably uniform storm for which reliable rainfall and runoff data are available.

2. Separation of the runoff hydrograph into direct surface runoff and base flow (ground water contribution). Determination of the direct surface runoff hydrograph by subtracting the base flow from the runoff hydrograph.

3. Determination of the volume of the direct surface runoff by planimetry (written in cubic meters net runoff).

4. Determination of the ordinates of the unit hydrograph by dividing the ordinates of the direct runoff hydrograph by the volume of the direct runoff. The resulting unit hydrograph should represent a unit volume runoff from the drainage area.

Example 10.2

Stream flow data were recorded for a storm whose effective duration was 12 h (see Table 10.1). The drainage area to the site of the measurements is 1150 km². Construct a unit hydrograph for this 12-h duration storm. (See Figure 10.10.)

TABLE 10.1

Day	(1) Hour	(2) Discharge	(3) Estimated Base Flow	(4) Direct Runoff	(5) Hours After Start
22	0600	150	150	0	0
	1200	400	157	243	6
	1800	750	165	585	12
23	0000	980	172	808	18
	0600	890	180	710	24
	1200	690	187	503	30
	1800	480	195	285	36
24	0000	370	202	168	42
	0600	300	210	90	48
	1200	260	217	43	54
	1800	225	225	0	60
25	0000	200	200	—	—
	0600	180	180	—	—
	1200	170	170	—	—

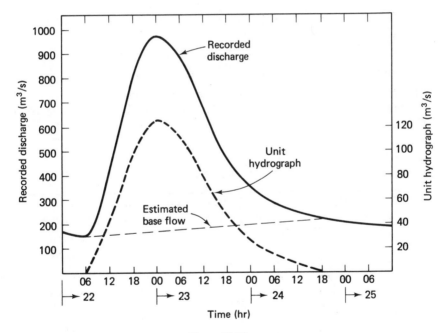

Figure 10.10

Solution

With the data provided, the step computations are carried out to obtain the coordinates for the unit hydrograph. The values in each column in Table 10.1 are explained as follows on page 320:

Column (1): Time (in hours) since the rainfall began.

Column (2): Recorded discharge, in cubic meters per second.

Column (3): Estimated base flow, determined as in Figure 10.8 in cubic meters per second.

Column (4): Direct runoff, obtained by subtracting column (3) from column (2), in cubic meters per second.

Column (5): Unit hydrograph ordinate found by dividing column (4) by the volume of the direct runoff (expressed in centimeters). Such a volume results from the summation of all values in column (4) divided by the drainage area.

$$h = \frac{3435 \frac{m^3}{s} \cdot 6h \cdot 3600 \frac{s}{h} \cdot \left(100 \frac{cm}{m}\right)^3}{1150 \text{ km}^2 \cdot \left(10^5 \frac{cm}{km}\right)^2}$$

$h = 6.45$ cm

Example 10.3

The basin in Example 10.2 recorded a storm that produced a 5-cm runoff in the first 12 h and a 3-cm runoff for the subsequent 12 h (see Table 10.2). Calculate the hydrograph expected in this storm.

TABLE 10.2

(1) Hour	(2) 5·Unit Hydrograph	(3) 3·Unit Hydrograph	(4) Estimated Base Flow	(5) Total Discharge
0	0	—	190	190.0
6	188.5	—	198	386.5
12	453.5	0	206	659.5
18	626.5	113.1	214	953.6
24	550.5	272.1	222	1044.6
30	390.0	375.9	230	995.9
36	221.0	330.3	238	789.3
42	130.0	234.0	246	610.0
48	70.0	132.6	254	456.0
54	33.5	78.0	262	373.5
60	0	42.0	270	312.0
66	—	20.1	278	298.1
72	—	0	286	286.0

Solution

The values in each column in Table 10.2 are explained as follows:

Column (1): The time since the storm began.

Column (2): Discharge related to the first 12 h of storm beginning at time zero.

Column (3): Discharge related to the second 12 h of storm beginning at time 12 h.

Column (4): Estimated base flow from conditions existing before the storm began.

Column (5): Total discharge obtained by adding columns (2), (3), and (4). The values in this column are plotted to give the expected hydrograph (see Figure 10.11).

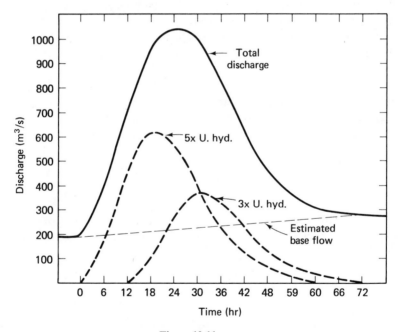

Figure 10.11

PROBLEMS

10.3.1. The following data were recorded at a river section during a flood. The data were collected at 6-h intervals. Compute the volume of direct runoff and the base flow.

Time (h)	Discharge (m³/s)	Time (h)	Discharge (m³/s)	Time (h)	Discharge (m³/s)
6	100	36	251	66	90
12	1490	42	190	72	77
18	1090	48	141	78	66
24	610	54	122	84	57
30	401	60	104	90	54

10.3.2. Select a flood record from a stream in your area from a copy of a U.S. Geological Survey Water-Suppply Paper. Construct the hydrograph and estimate the corresponding direct runoff and base flow volumes.

10.4   FREQUENCY ANALYSIS

Design of hydraulic structures frequently requires information for estimating the probability of future occurrence of certain hydrological events, such as the magnitude of flood, the intensity of precipitation, the period of continuous drought, etc. Frequency analysis provides information for estimating the probability of future occurrence based on past data of a particular event recorded.

In computing the data for frequency analysis, an order (m) is assigned to each individually recorded event according to the magnitude ($m = 1$ for the highest magnitude in the series). The probability of an event of the same magnitude, or higher, occurring may be calculated by

$$P = \frac{m}{N+1} \tag{10.1}$$

where N is the total number of events recorded in the data series. The calculated frequency of an event of the same magnitude or higher reoccurring is

$$T_r = \frac{1}{P} \tag{10.2}$$

The following example demonstrates the basic procedure of frequency analysis.

Example 10.4

A continuous record of annual maximum flood discharge (Table 10.3) was taken at New London, Wisconsin, for the Wolf River (tributary to Lake Michigan) between 1896 and 1968, a total of 73 years. Frequency analysis was performed on the data series for the annual peak discharge.

TABLE 10.3 Peak Discharges of the Wolf River
at New London, Wisconsin

(1) Year	Discharge		(4) Order (m)	(5) P	(6) T_r (yr)
	(2) (CFS)	(3) (m^3/s)			
1896	3420	96.79	64	0.865	1.16
1897	4390	124.24	55	0.743	1.35
1898	2865	81.08	71	0.959	1.04
1899	5430	153.67	47	0.635	1.57
1900	2750	77.83	72	0.973	1.03
1901	6230	176.31	40	0.540	1.85
1902	3050	86.32	69	0.932	1.07
1903	5100	144.33	52	0.703	1.42
1904	5160	146.03	50	0.676	1.48
1905	6470	183.10	36	0.486	2.06
1906	7250	205.18	32	0.432	2.31
1907	5100	144.33	51	0.689	1.45
1908	4350	123.11	56	0.757	1.32
1909	3420	96.79	65	0.878	1.14
1910	3500	99.05	62	0.838	1.19
1911	3120	88.30	68	0.919	1.09
1912	9180	259.79	16	0.216	4.63
1913	8170	231.21	20	0.270	3.70
1914	8490	240.27	18	0.243	4.11
1915	4260	120.56	58	0.784	1.28
1916	8960	253.57	17	0.230	4.35
1917	8060	228.10	21	0.284	3.52
1918	7270	205.74	31	0.419	2.39
1919	6350	179.71	38	0.514	1.95
1920	10,800	305.64	8	0.108	9.25
1921	6560	185.65	35	0.473	2.11
1922	15,500	438.65	1	0.014	74.00
1923	10,100	285.83	13	0.176	5.69
1924	7280	206.02	30	0.405	2.47
1925	4270	120.84	57	0.770	1.30
1926	4470	126.50	54	0.730	1.37
1927	6340	179.42	39	0.527	1.90
1928	7810	221.02	24	0.324	3.08
1929	11,300	319.79	6	0.081	12.33
1930	2900	82.07	70	0.946	1.06
1931	2160	61.13	73	0.986	1.01
1932	4260	120.56	59	0.797	1.25
1933	5320	150.56	48	0.649	1.54
1934	6000	169.80	43	0.581	1.72
1935	9570	270.83	15	0.203	4.93
1936	7450	210.84	28	0.378	2.64
1937	6360	179.99	37	0.500	2.00
1938	11,500	325.45	5	0.068	14.80
1939	11,100	314.13	7	0.095	10.57

TABLE 10.3 Cont.

| | Discharge | | (4) | | |
(1) Year	(2) (CFS)	(3) (m³/s)	Order (m)	(5) P	(6) Tᵣ (yr)
1940	4880	138.10	53	0.716	1.40
1941	7140	202.06	33	0.446	2.24
1942	7940	224.70	22	0.297	3.36
1943	11,700	331.11	4	0.054	18.50
1944	6080	172.06	42	0.568	1.76
1945	7600	215.08	26	0.351	2.85
1946	10,300	291.49	11	0.149	6.73
1947	5970	168.95	44	0.595	1.68
1948	5460	154.52	46	0.622	1.61
1949	4020	113.77	60	0.811	1.23
1950	7000	198.10	34	0.459	2.18
1951	10,500	297.15	9	0.122	8.22
1952	15,200	430.16	2	0.027	37.00
1953	10,400	294.32	10	0.135	7.40
1954	3980	112.63	61	0.824	1.21
1955	5830	164.99	45	0.608	1.64
1956	7470	211.40	27	0.365	2.74
1957	3320	93.96	66	0.892	1.12
1958	3210	90.84	67	0.905	1.10
1959	7840	221.87	23	0.311	3.22
1960	13,300	376.39	3	0.041	24.67
1961	7430	210.27	29	0.392	2.55
1962	8490	240.27	19	0.257	3.89
1963	7700	217.91	25	0.338	2.96
1964	3450	97.64	63	0.851	1.17
1965	9990	282.72	14	0.189	5.29
1966	5190	146.88	49	0.662	1.51
1967	10,200	288.66	12	0.162	6.17
1968	6170	174.61	41	0.554	1.80

Solution

Drainage area: 5800 km² (2240 sq. miles)

Gage: Staff gage until 1951, graphic water-stage gage until 1964, and digital water-stage gage subsequently.

Source: U.S.G.S. Open File Report I 19.2: W75 (1971).

$$P = \frac{m}{N+1} \qquad T_r = \frac{1}{P}$$

$$N = 73$$

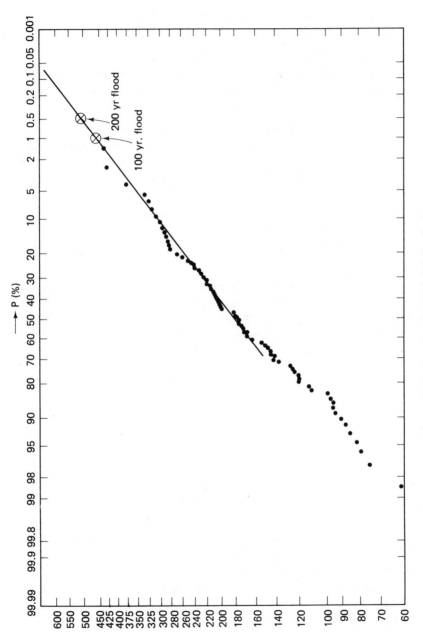

Figure 10.12. Log-Pearson Type III distribution.

The values in the columns in Table 10.3 on pages 323 and 324 are explained as follows:

Column (1): Years listed in chronological order.

Column (2): Annual peak discharge recorded for Wolf River at New London, Wisconsin (in cubic feet per second).

Column (3): Peak discharge rates converted to cubic meters per second.

Column (4): Order (m) assigned to each recorded peak discharge according to the magnitude.

Column (5): Calculated probability of an event with equal or higher magnitude occurring in any given year. For example, the probability of a flood with peak discharge equal to or greater than 331 m³/s occurring in any given year is approximately 5.4%

Column (6): Calculated return period. For example, the return period of a flood with peak discharge equal to or greater than 331 m³/s is 18½ years.

The values in columns (3) and (5) are plotted on a probability paper as shown in Figure 10.12. The line drawn through the data points can be extrapolated to estimate the magnitude of the 100-yr flood or the 200-yr flood, if necessary. The corresponding probabilities are:

$$P(100) = \frac{1}{100 + 1} = 0.00990$$

$$P(200) = \frac{1}{200 + 1} = 0.00498$$

PROBLEMS

10.4.1. Assume that the curve in Figure 10.12 represents the true average return period of an annual flood. What is the discharge of an annual flood that would only occur once in 20 years?

10.4.2. Compile an annual series of minimum flow for a selected gaging station on a stream in your area (from U.S.G.S. Water-Supply Papers) and plot the frequency curve on Log-Pearson type III paper.

10.4.3. Obtain excessive precipitation data for a selected first-order U.S. Weather Bureau station (from Bulletin W) and derive the intensity-frequency curves for durations of 15 min, 30 min, 60 min, and 90 min.

10.5 FLOOD ROUTING IN OPEN CHANNELS

In engineering hydrology it is frequently necessary to predict the discharge hydrograph of a flood at a downstream site based on the discharge hydrograph obtained from an upstream station in the same stream channel. The procedure involved in such a flood prediction is known as *flood routing*.

Flood routing may be applied to predict the flood discharge or stage at any downstream station in a stream channel. It may also be used as a tool for analyzing the function and operation of an outlet structure in a proposed reservoir. The former application is commonly known as *stream routing*.

In engineering practice the shape of the hydrograph curve is of great importance. Generally speaking, the downstream hydrograph is different from the upstream hydrograph in two ways: (1) The downstream hydrograph has a lower peak and a broad base and (2) the peak occurs at a later time. The lower peak tells the degree of peak flow reduction resulting from the passage of the flood wave through the reach of the stream channel. The broader base tells the duration of the higher than normal water stage (or discharge) that might threaten the dikes or similar structures to control the flood. The delay of the peak is important in determiming whether sufficient time is provided to avoid simultaneous occurrence of flood peaks from the tributaries, for example.

Flood routing in a given stream channel is based on the simple relation of continuity between the two stations at the upstream and downstream ends of the channel reach.

$$I \cdot \Delta T = O \cdot \Delta T + S$$

where I is the inflow, O is the outflow from the reach in unit time, and S is the change in storage in the reach during the time period ΔT. In the passage of a flood wave all three quantities are varying with time. If the increments of time ΔT selected are sufficiently short, the inflow and outflow hydrograph can be approximated by successive segments of straight lines, and the above equation may be rewritten as

$$\left(\frac{I_1 + I_2}{2}\right)\Delta T = \left(\frac{O_1 + O_2}{2}\right)\Delta T + (S_2 - S_1) \tag{10.3}$$

where the subscripts 1 and 2 represent the beginning and the end of an increment of time.

In a typical flood-routing problem the inflow hydrograph is always known along with the initial values of outflow, O_1 and storage, S_1. Rearranging the equation gives

$$S_2 + \frac{O_2}{2}\Delta T = \left(\frac{I_1 + I_2}{2}\right)\Delta T + S_1 - \frac{O_1}{2}\Delta T \tag{10.4}$$

The quantities on the left-hand side of the equation are the unknowns. In order to solve the above equation for the two unknowns, an independent relationship between the outflow O and storage S is needed. In stream routing, the storage in a reach of the stream channel may be expressed as

$$S = KO + K(I - O) \tag{10.5}$$

The first term on the right-hand side of the above equation is known as the *prism storage*. It represents the volume of water that would be stored in the reach if the discharge were uniform throughout the reach. K has the dimension of time and is interpreted as the time of travel through the channel reach. The second term is called the *wedge storage*. It represents the increment of storage, in addition to the prism storage, resulting from the difference in inflow to the reach and outflow from the reach at a particular time. The flow rate increase in a channel due to a storm is usually not a linear function of time, as is schematically shown in Figure 10.13. During any particular

Figure 10.13

time period K the wedge storage can be represented by the shaded area, which is a portion of $K(I - O)$. This portion is given by the percentage factor, ϕ, which is always less than unity.

Substituting Eq. (10.5) into Eq. (10.4) and rearranging the terms give

$$O_2 = C_1 I_1 + C_2 I_2 + C_3 O_1 \tag{10.6}$$

where

$$C_1 = \frac{K\phi + \frac{1}{2}\Delta T}{K - K\phi + \frac{1}{2}\Delta T}$$

$$C_2 = -\frac{K\phi - \frac{1}{2}\Delta T}{K - K\phi + \frac{1}{2}\Delta T}$$

$$C_3 = \frac{K - K\phi - \frac{1}{2}\Delta T}{K - K\phi + \frac{1}{2}\Delta T}$$

Given the inflow hydrograph and the initial values of the outflow, Eq. (10.6) can be used to calculate the outflow hydrograph.

To determine the values of K and ϕ for a channel reach, field measurements must be made to obtain data of inflow and corresponding outflow. Rearranging Eq. (10.5) gives

$$S = K[\phi I + (1 - \phi)O] \qquad (10.7)$$

The values of channel storage, S, can be determined by

$$S = \Sigma(I - O)\Delta T \qquad (10.8)$$

By measuring the corresponding inflow and outflow in the field for some existing flood conditions, the values of Eq. (10.8) can be plotted against those of Eq. (10.7) with assumed values of ϕ. The value of ϕ is selected by trial and error so that the data can be plotted as a single-value straight line on linear scales. K is the slope of the straight line, as shown in the following example.

Example 10.5

Determine K and ϕ values for a reach of river. Field measurements are shown in columns (2) and (3) in Table 10.4.

TABLE 10.4

(1) Time (h)	(2) Inflow (m^3/s)	(3) Outflow (m^3/s)	(4) ΔS (m^3/s /day)	(5) S (m^3/s /day)	(6) $\phi I + (1 - \phi)O$		
					$\phi = 0.2$	$\phi = 0.25$	$\phi = 0.3$
0	40	40	0	0	40	40	40
6	160	52	27	27	73.6	79	84.4
12	382	60	80.5	107.5	124.4	140.5	156.6
18	548.8	124	106.2	213.7	209.0	230.2	251.4
24	498	240.8	64.3	278	292.2	305.1	318
30	408	316	23	301	334.4	339	343.6
36	328	352	−6	295	347.2	346	344.8
42	260	348	−22	273	330.4	326	321.6
48	220	328	−27	246	306.4	301	295.6
54	188	300	−28	218	277.6	272	266.4
60	164	270	−26.5	191.5	248.8	243.5	238.2
66	144	246	−25.5	166	225.6	220.5	215.4
72	124	232	−27	139	210.4	205	199.6
78	108	204	−24	115	184.8	180	175.2
84	97	180	−21	94	163.2	159	154.8
90	84	156	−18	76	141.6	138	134.4

Solution

The values in the columns in Table 10.4 are explained as follows:

Column (1): Time since the beginning of the inflow wave (in hours).

Columns (2) and (3): Inflow and outflow field measurements (in cubic meters per second).

Column (4): Change in storage (in cubic meters per second per day).
$$\Delta S = (I - O) \cdot 6 \, \text{h} \cdot \frac{1 \, \text{day}}{24 \, \text{h}}$$

Column (5): Cumulative change in storage (in cubic meters per second per day).

Column (6): Right-hand side variable in Eq. (10.7). The value that produces the nearest single-valued straight line is the chosen one (Figure 10.14). This column is given in cubic meters per second. The slope of the line represents K. The chosen value is $\phi = 0.3$, then, $K = 1.06$.

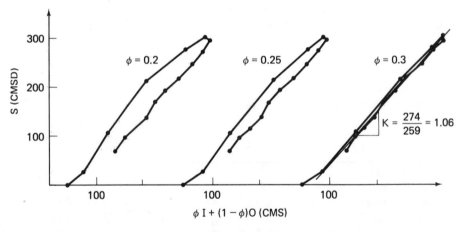

Figure 10.14

Example 10.6

With the K and ϕ values obtained in Example 10.5, calculate the outflow hydrograph for the channel reach with the inflow hydrograph provided in Table 10.5. Compare the inflow and outflow hydrographs.

Solution

From Figure 10.14, we have,

$$K = 1.06; \qquad \phi = 0.3; \qquad \Delta T = 12 \, \text{h} = 0.5 \, \text{days}$$

The constants of Eq. (10.6) are:

$$C_1 = \frac{K\phi + \frac{1}{2}\Delta T}{K - K\phi + \frac{1}{2}\Delta T} = \frac{1.06 \cdot 0.3 + \frac{1}{2} \cdot 0.5}{1.06 - 1.06 \cdot 0.3 + \frac{1}{2} \cdot 0.5} = \frac{0.568}{0.992} = 0.573$$

$$C_2 = -\frac{K\phi - \frac{1}{2}\Delta T}{K - K\phi + \frac{1}{2}\Delta T} = \frac{1.06 \cdot 0.3 - \frac{1}{2} \cdot 0.5}{0.992} = -\frac{0.068}{0.992} = -0.069$$

$$C_3 = \frac{K - K\phi - \frac{1}{2}\Delta T}{K - K\phi + \frac{1}{2}\Delta T} = \frac{1.06 - 1.06 \cdot 0.3 - \frac{1}{2} \cdot 0.5}{0.992} = \frac{0.492}{0.992} = 0.496$$

The computations are carried out in the following order as listed in Table 10.5. The results are plotted in Figure 10.15.

TABLE 10.5

(1) Time (h)	(2) Inflow (m³/s)	(3) $C_1 I_1$	(4) $C_2 I_2$	(5) $C_3 O_1$	(6) O
0	48	—	—	—	48
12	93	27.5	−6.4	23.8	44.9
24	288	53.3	−19.9	22.3	55.7
36	366	165.0	−25.3	27.6	167.3
48	306	209.2	−21.1	83.0	271.6
60	255	175.3	−17.6	134.7	292.4
72	210	146.1	−14.5	145.0	276.6
84	171	120.3	−11.8	137.2	245.7
96	141	98.0	−9.7	121.9	210.2
108	117	80.8	−8.1	104.3	177.0
120	96	67.0	−6.6	87.8	148.2
132	84	55.0	−5.8	73.5	122.7
144	72	48.1	−5.0	60.9	104.0
156	66	41.3	−4.6	51.6	88.3
168	60	37.8	−4.1	41.4	75.1
180	57	34.4	−3.9	35.2	65.7

Following this procedure, I_1 and O_1 are taken from the preceding line, while I_2 is from the same line being worked.

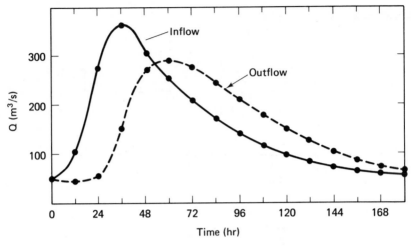

Figure 10.15

COMMON CONSTANTS AND CONVERSIONS

At normal conditions (20.2°C and 760 mm Hg) water properties are:

- Specific weight   = 9789 N/m^3 $\equiv$ 62.32 lb/ft^3
- Density   = 998.2 kg/m^3 $\equiv$ 1.94 slug/ft^3
- Viscosity   = $1 \cdot 10^{-3}$ N·s/m^2 $\equiv$ $2.089 \cdot 10^{-5}$ lb·s/ft^2
- Kinematic viscosity = $1.002 \cdot 10^{-6}$ m^2/s $\equiv$ $1.078 \cdot 10^{-5}$ ft^2/s

At standard conditions (4°C and 760 mm Hg) water properties are:

- Specific weight   = 9806 N/m^3 $\equiv$ 62.43 lb/ft^3
- Density   = 1000 kg/m^3 $\equiv$ 1.94 slug/ft^3
- Viscosity   = $1.57 \cdot 10^{-3}$ N·s/m^2 $\equiv$ $3.28 \cdot 10^{-5}$ lb·s/ft^2
- Kinematic viscosity = $1.57 \cdot 10^{-6}$ m^2/s $\equiv$ $1.69 \cdot 10^{-5}$ ft^2/s

Standard atmospheric pressure = 101,376 N/m^2

Gravitational constant = 9.81 m/s^2

Length

$$\frac{0.3048 \text{ m}}{1 \text{ ft}} = 1 \qquad \frac{2.54 \text{ cm}}{1 \text{ in.}} = 1 \qquad \frac{100 \text{ cm}}{1 \text{ m}} = 1$$

Area

$$\frac{9.29 \cdot 10^{-2} \text{ m}^2}{1 \text{ ft}^2} = 1 \qquad \frac{6.452 \text{ cm}^2}{1 \text{ in.}^2} = 1 \qquad \frac{4046.86 \text{ m}^2}{1 \text{ acre}} = 1$$

$$\frac{640 \text{ acre}}{1 \text{ mile}^2} = 1 \qquad \frac{1 \cdot 10^6 \text{ m}^2}{1 \text{ km}^2} = 1 \qquad \frac{1 \cdot 10^4 \text{ cm}^2}{1 \text{ m}^2} = 1 \qquad \frac{2.59 \text{ km}^2}{1 \text{ mile}^2} = 1$$

Volume

$$\frac{2.832 \cdot 10^{-2} \text{ m}^3}{1 \text{ ft}^3} = 1 \qquad \frac{1 \cdot 10^6 \text{ cm}^3}{1 \text{ m}^3} = 1 \qquad \frac{1 \cdot 10^3 \text{ l}}{1 \text{ m}^3} = 1$$

$$\frac{7.4805 \text{ gal}}{1 \text{ ft}^3} = 1 \qquad \frac{3.7854 \text{ l}}{1 \text{ gal}} = 1 \qquad \frac{43560 \text{ ft}^3}{1 \text{ acre-ft}} = 1$$

$$\frac{3.2585 \cdot 10^5 \text{ gal}}{1 \text{ acre-ft}} = 1 \qquad \frac{1233.48 \text{ m}^3}{1 \text{ acre-ft}} = 1$$

Velocity

$$\frac{0.3048 \text{ m/s}}{1 \text{ ft/s}} = 1 \qquad \frac{30.48 \text{ cm/s}}{1 \text{ ft/s}} = 1$$

Discharge

$$\frac{694.44 \text{ GPM}}{1 \text{ MGD}} = 1 \qquad \frac{448.83 \text{ GPM}}{1 \text{ ft}^3/\text{s}} = 1 \qquad \frac{226.29 \text{ GPM}}{1 \text{ acre-ft/day}} = 1$$

$$\frac{15848.5 \text{ GPM}}{1 \text{ m}^3/\text{s}} = 1 \qquad \frac{86400 \text{ m}^3/\text{day}}{1 \text{ m}^3/\text{s}} = 1 \qquad \frac{3153.6 \text{ cm} \cdot \text{km}^2/\text{year}}{1 \text{ m}^3/\text{s}} = 1$$

$$\frac{479.3 \text{ in./year/mile}^2}{1 \text{ m}^3/\text{s}} = 1$$

Mass

$$\frac{14.594 \text{ kg}}{1 \text{ slug}} = 1 \qquad \frac{14594 \text{ g}}{1 \text{ slug}} = 1$$

Density

$$\frac{515.4 \text{ kg/m}^3}{1 \text{ slug/ft}^3} = 1 \qquad \frac{0.5154 \text{ g/cm}^3}{1 \text{ slug/ft}^3} = 1$$

Force

$$\frac{4.4482 \text{ N}}{1 \text{ lb}} = 1 \qquad \frac{1 \cdot 10^5 \text{ dyn}}{1 \text{ N}} = 1$$

Pressure

$$\frac{6894.7 \text{ N/m}^2}{1 \text{ psi}} = 1 \qquad \frac{68947 \text{ dyn/cm}^2}{1 \text{ psi}} = 1 \qquad \frac{6.8011 \cdot 10^{-2} \text{ atm}}{1 \text{ psi}} = 1$$

$$\frac{1 \text{ bar}}{1 \cdot 10^6 \text{ dyn/cm}^2} = 1$$

Temperature

$$(°F) = 1.8(°C) + 32, \text{ i.e., } 20°C \equiv 68°F$$

Energy

$$\frac{1 \text{ Joule}}{1 \text{ N} \cdot \text{m}} = 1 \qquad \frac{1 \cdot 10^7 \text{ erg}}{1 \text{ Joule}} = 1 \qquad \frac{1 \text{ erg}}{1 \text{ dyn} \cdot \text{cm}} = 1 \qquad \frac{0.23901 \text{ cal}}{1 \text{ Joule}} = 1$$

Viscosity

$$\frac{10 \text{ Poise}}{1 \text{ N} \cdot \text{s/m}^2} = 1 \qquad \frac{42.88 \text{ N} \cdot \text{s/m}^2}{1 \text{ lb} \cdot \text{s/ft}^2} = 1 \qquad \frac{100 \text{ cP}}{1 \text{ Poise}} = 1$$

Kinematic Viscosity

$$\frac{1 \cdot 10^4 \text{ Stoke}}{1 \text{ m}^2/\text{s}} = 1 \qquad \frac{9.294 \cdot 10^{-2} \text{ m}^2/\text{s}}{1 \text{ ft}^2/\text{s}} = 1 \qquad \frac{1 \text{ cm}^2/\text{s}}{1 \text{ Stoke}} = 1$$

$$\frac{100 \text{ cS}}{1 \text{ Stoke}} = 1$$

Power

$$\frac{550 \text{ lb} \cdot \text{ft/s}}{1 \text{ hp}} = 1 \qquad \frac{745.7 \text{ N} \cdot \text{m/s}}{1 \text{ hp}} = 1 \qquad \frac{1.356 \text{ N} \cdot \text{m/s}}{1 \text{ lb} \cdot \text{ft/s}} = 1$$

$$\frac{1 \text{ W}}{1 \text{ Joule/s}} = 1 \qquad \frac{1 \text{ W}}{1 \text{ N} \cdot \text{m/s}} = 1$$

A

GRAPHICAL FLOW NETS: ELECTRIC ANALOG AND NUMERICAL ANALYSIS

Two dimension flow nets are constructed with sets of streamlines (ψ lines) and equipotential lines (ϕ-lines). The streamlines are tangential to the velocity vector at all points in the flow field, while the equipotential lines represent constant velocity potential levels. Usually, the streamlines are chosen so as to divide the flow into a number of channels conveying equal flow rates, and the spacing of the equipotential lines is so selected that it is equal to the spacing between the streamlines everywhere in the flow field.

Some general characteristics of flow nets can be listed as follows:

1. Flow nets represent the flow patterns of irrotational flow* within given boundaries. It is not necessary that the flow be steady.

2. There is only one possible pattern of flow for a given set of boundary conditions.

3. The ϕ-lines intersect the ψ-lines and the fixed boundaries, at 90°.

4. Each unit grid resulting from the ψ-lines and ϕ-lines approximate to squares, having equal median lines and 90° angles (except at points of zero velocity or boundary discontinuity).

5. In regions of uniform flow the squares are of equal size. In diverging flows they increase, and in converging flows they decrease in size in the

*Flow of zero vorticity

direction of the flow. The size of the square is inversely proportional to the velocity at any point in the flow field.

The graphical procedure for the construction of a flow net may be summarized in the following steps:

1. Decide upon the number of streamlines required to yield the necessary details of the flow pattern. Sketch the streamlines in regions where the velocity distributions are evident, such as in parallel or radial flow to a well.

2. Sketch lightly the remaining portions of the streamlines with smooth curves. The spacing should decrease with decrease of radius on curves. The product of velocity and the radius is always constant in irrotational flows.

3. Sketch the equipotential lines; they should normally intersect all streamlines including the boundaries, and form squares with the streamlines. It will be found necessary to adjust the initial streamline locations from time to time until the above conditions are satisfied. The process is a trial-and-error one, in which the operator improves the sketches with practice.

4. At the stagnation points (points of zero velocity), five-sided units occur. In this case, the equipotential lines adjacent to the stagnation point should be spaced nearly equally about the stagnation point.

5. Free water table surface must be located by trial such that, when the flow net is drawn, the velocity along the free surface is in accord with the boundary conditions of the problem. Since the free surface of the water table is affected by gravity, the spacing of the ϕ-lines along the water table should be such that the velocity head at any point on the water table equals the distance of the point below the total head line, h_0 (Figure A.1).

$$V = \sqrt{2gh_v}\,\frac{\Delta\phi}{\Delta s}$$

where $\Delta\phi$ is the spacing between the ϕ-lines and Δs is the spacing between the ψ-lines, h_v is the velocity head at the point

$$h_v = h_0 - h.$$

6. The diagonals of the "squares" should form smooth curves which intersect each other normally. This fact is used to check the accuracy of the final flow pattern.

Figure A.1. Gravity affected water table.

ELECTRIC ANALOG

The flow of an electric current in a two-dimensional conductor is analogous to irrotational flow. The electrical potential (voltage) is the counterpart of the velocity potential, and the electric current (I) is analogous to the fluid flow. A sheet of uniform, homogeneous conducting material or a shallow bath of electrolyte of uniform depth is shaped so that its boundary is geometrically similar to that of the prototype flow field to be investigated (for example, Figures A.2a and A.2b). If a voltage drop is established along the

(a) Equipotential lines   (b) Streamlines

Figure A.2. Electrode arrangements in an electrical analog model for determining (a) equipotential lines, and (b) streamlines of seepage beneath a masonry dam.

conductor representing the boundary equipotential lines, a potential pattern geometrically similar to that in the flow field is traced (Figure A.2a). The electric potential p is the counterpart of the velocity potential function ϕ, and the components of the electric field strength, E_x and E_y, are the counterparts of velocity components V_x and V_y, respectively. The corresponding equations for the electric field are

$$\frac{\partial^2 p}{\partial x^2} + \frac{\partial^2 p}{\partial y^2} = 0 \tag{1}$$

and

$$E_x = \frac{\partial p}{\partial x}, \qquad E_y = \frac{\partial p}{\partial y} \tag{2}$$

If a voltage drop is established along the boundary streamlines, then the ψ-line pattern can be obtained (Figure A.2b).

A voltmeter or potentiometer probe is used to locate the lines of constant potential. By fixing a constant voltage drop between the boundary voltage, each ϕ-line and ψ-line can be accurately located. The electrical analogy arrangement is shown in Figure A.2.

NUMERICAL ANALYSIS

Since the streamlines are everywhere parallel to the velocity vector in the flow field, a stream function, ψ, may be defined such that

$$\frac{\partial \psi}{\partial y} = V_x, \qquad \frac{\partial \psi}{\partial x} = -V_y \tag{3}$$

where V_x and V_y are the velocity components in the x- and y-directions respectively.

In a flow field free from vortex motion, or where the effect of vortices can be neglected, the flow is said to be *irrotational*. Flow net may be constructed in irrotational flow fields where the *rotation of flow* is zero. The rotation of flow is defined as

$$\omega = \frac{1}{2}\left(\frac{\partial V_y}{\partial x} - \frac{\partial V_x}{\partial y}\right)$$

For irrotational flow, then

$$\frac{\partial V_y}{\partial x} - \frac{\partial V_x}{\partial y} = 0 \tag{4}$$

Substituting Equations (3) into Equation (4) and simplifying, we have,

$$\frac{\partial^2 \psi}{\partial x^2} + \frac{\partial^2 \psi}{\partial y^2} = 0 \tag{5}$$

which is commonly known as the *Laplace Equation*.

Similarly, a velocity potential function, ϕ, may be defined for the irrotational flow field such that the gradient of ϕ gives the velocity component in each respective direction.

$$\frac{\partial \phi}{\partial x} = V_x \qquad \frac{\partial \phi}{\partial y} = V_y \qquad (6)$$

In the meantime, the continuity condition requires that,

$$\frac{\partial V_x}{\partial x} + \frac{\partial V_y}{\partial y} = 0 \qquad (7)$$

Substituting Eq. (6) into Eq. (7) and simplifying, we have

$$\frac{\partial^2 \phi}{\partial x^2} + \frac{\partial^2 \phi}{\partial y^2} = 0 \qquad (8)$$

A streamline (ψ-line) is defined as a line for which the stream function ψ has a constant value. An *equipotential line* (ϕ-line) is defined as a line for which the velocity potential function ϕ has a constant value. Flow net is constructed with a set of streamlines and its corresponding set of equipotential lines.

Impervious boundaries are streamlines since flow velocity is parallel and no flow is allowed to cross the boundary lines. In a certain flow field (for example, Fig. A.4), an arbitrary value may be assigned to each boundary stream function. Say, one boundary has the stream function value of 0 and the other boundary has the stream function value of 100. Eq. (5) may be solved by numerical integration in a quadrangular grid (for example, Fig. A.4b) for given boundary conditions. The discretization of the Laplace equation may be expressed as follows:

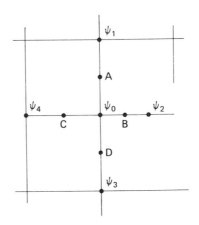

Figure A-3

A ψ-value is first assigned to each node in the quadrangular grid (Fig. A.3), and the gradient of ψ at the mid-point of each grid arm (A, B, C, and D) may be expressed as, respectively,

$$\frac{\partial \psi}{\partial y}\bigg|_A = \frac{\psi_1 - \psi_0}{L_1} \tag{9}$$

$$\frac{\partial \psi}{\partial x}\bigg|_B = \frac{\psi_2 - \psi_0}{L_2} \tag{10}$$

$$\frac{\partial \psi}{\partial y}\bigg|_C = \frac{\psi_0 - \psi_3}{L_3} \tag{11}$$

$$\frac{\partial \psi}{\partial x}\bigg|_D = \frac{\psi_0 - \psi_4}{L_4} \tag{12}$$

where the L's are the length of each arm measured between the respective nodes. The arms may or may not have the same length depending on the boundary geometry. Combining Eqs. (9) and (11), we may write the second derivative of ψ at the node 0:

$$\frac{\partial^2 \psi}{\partial y^2}\bigg|_0 = \frac{\dfrac{2\psi}{2y}\bigg|_A - \dfrac{2\psi}{2y}\bigg|_C}{\left(\dfrac{L_1 + L_3}{2}\right)} = \frac{2\left(\dfrac{\psi_1}{L_1} + \dfrac{\psi_3}{L_3}\right)}{L_1 + L_3} - \frac{2\psi_0}{L_1 \cdot L_3} \tag{13}$$

Using the same approach, we may also write,

$$\frac{\partial^2 \psi}{\partial x^2}\bigg|_0 = \frac{2\left(\dfrac{\psi_2}{L_2} + \dfrac{\psi_4}{L_4}\right)}{L_2 + L_4} - \frac{2\psi_0}{L_2 \cdot L_4} \tag{14}$$

Substituting Eqs. (13) and (14) into the Laplace equation (Eq. 5), we have:

$$\frac{\dfrac{\psi_2}{L_2} + \dfrac{\psi_4}{L_4}}{L_2 + L_4} + \frac{\dfrac{\psi_1}{L_1} + \dfrac{\psi_3}{L_3}}{L_1 + L_3} - \frac{\psi_0}{L_2 \cdot L_4} - \frac{\psi_0}{L_1 \cdot L_3} = 0 \tag{15}$$

or,

$$\psi_0\left(\frac{1}{L_2 \cdot L_4} + \frac{1}{L_1 \cdot L_3}\right) = \frac{\dfrac{\psi_2}{L_2} + \dfrac{\psi_4}{L_4}}{L_2 + L_4} + \frac{\dfrac{\psi_1}{L_1} + \dfrac{\psi_3}{L_3}}{L_1 + L_3} \tag{16}$$

when the arms are of the same length, $L_1 = L_2 = L_3 = L_4 = a$, which is the length of the basic grid selected, the above relationship may be simplified to

$$\psi_0 = \frac{a^2}{2}\left(\frac{\psi_2 + \psi_4}{2a^2} + \frac{\psi_1 + \psi_3}{2a^2}\right) = \frac{\psi_1 + \psi_2 + \psi_3 + \psi_4}{4} \tag{17}$$

When a digital computer is used, a grid size is first adopted and the following information included: the initial value of stream function for every node in the mesh, the length of each arm according to the boundary geometry given, and a code specifying if a node is on a boundary or if it has one or more than one arms shorter than the basic grid size a. Generally speaking, the smaller the grid the more accurate the solution. However, the resolution is always limited by the capacity of the computer and the cost of operation.

Equipotential lines may be constructed in exactly the same way using assumed values for equipotential lines far upstream and downstream from the object.

Example 1.A

Determine the pattern of streamlines for the seepage through the foundation of a dam as shown in Figure A.4.

(a)

(b)

Figure A-4

Solution:

A basic grid with arms length of 8 m appears to fit the boundary geometry fairly well. The dam bottom surface, as a boundary, is assigned a stream function value of 0; and the impervious layer deep in the foundation is assigned the value of 100.

Far from the dam, the direction of the seepage flow is assumed small and parallel to the horizontal direction. At these stations, same value as that of the impervious boundary is assumed. The effect of these assumptions diminishes with distance. It becomes negligible near the base of the dam.

To completely define the boundaries, values are also assigned to the water-foundation interface. It is very unlikely that the right values can be guessed at the first trial. Since the streamlines must intersect the boundary at right angles, we also have for the interface,

$$\frac{\partial \psi}{\partial y} = 0, \qquad \text{or,} \qquad \frac{\psi_1 - \psi_0}{L_1} = 0, \qquad \text{then,} \qquad \psi_1 = \psi_0$$

A computer program is written to solve the Laplace equation of stream function and the result is plotted in Figure A.5 with the streamlines proportioned in as shown.

If it is desired to have more detailed flow patterns for a particular region, the computer program may be applied again to the region of interest using smaller grid and boundary conditions obtained from the previous computation. For example, the rectangular region bounded by ABCD in Fig. A.5 may be reprocessed by the same computer program using grid of one half the original size and the boundary conditions presented in the result of the above Example.

```
      C     NUMERICAL INTEGRATION OF THE LAPLACE EQUATION BY THE FINITE
      C     DIFFERENCES METHOD. A MESH IS READ IN, BOUNDARY CONDITIONS ARE
      C     INCLUDED IN THIS INFORMATION. THEN THE NODES ARE RELAXED AND THE
      C     FINAL RESULT PRINTED.
      C
      C
            DIMENSION NODE(6,15),K(6,15),HA(6,15),VA(6,15)
            REAL NODE
      C
      C   **READ MESH INFORMATION. K=0 FOR REGULAR NODES.  K=1 FOR BOUNDARY
      C     NODES.  AND K=2 FOR NODES WITH SHORTER ARMS.
      C
            READ(5,1)N,M,ITER,TOL
    1       FORMAT(3I5,F10.0)
            DO 2 I=1,N
    2       READ(5,3)(HA(I,J),J=1,M-1)
    3       FORMAT(16F5.0)
            DO 4 I=1,N-1
    4       READ(5,5)(VA(I,J),J=1,M)
    5       FORMAT(16F5.0)
            WRITE(6,6)TOL
    6       FORMAT(//////5X,'RELAXATION OF A FLOW NET FOR A GIVEN SET OF BOU
           1NDARY CONDITIONS'////5X,'NODES TOLERANCE=',F5.2//////5X,'SIZE OF
           2MESH ARMS'///)
            DO 7 I=1,N-1
    7       WRITE(6,8)(HA(I,J),J=1,M-1),(VA(I,J),J=1,M)
    8       FORMAT(4X,14F8.2//15F8.2//)
            WRITE(6,9)(HA(N,J),J=1,M-1)
    9       FORMAT(4X,14F8.2)
            DO 10 I=1,N
   10       READ(5,11)((K(I,J),NODE(I,J)),J=1,M)
   11       FORMAT(16(I1,F4.0))
            WRITE(6,12)
   12       FORMAT(//////5X,'INITIAL NODE INFORMATION MATRIX'////)
```

344

```fortran
      DO 13 I=1,N
   13 WRITE(6,14)(NODE(I,J),J=1,M)
   14 FORMAT(15F8.2//)
C
C     **RELAX NODES USING DISCRETE FORM OF THE LAPLACE EQUATION.
C
      DO 20 L=1,ITER
      IT=0
      DO 18 J=2,M-1
      DO 18 I=2,N-1
      TEST=NODE(I,J)
      IF(K(I,J)-1)15,18,16
   15 NODE(I,J)=(NODE(I-1,J)+NODE(I+1,J)+NODE(I,J-1)+NODE(I,J+1))/4.0
      GO TO 17
   16 C1=(NODE(I-1,J)/VA(I-1,J)+NODE(I+1,J)/VA(I,J))/(VA(I-1,J)+VA(I,J))
      C2=(NODE(I,J-1)/HA(I,J-1)+NODE(I,J+1)/HA(I,J))/(HA(I,J-1)+HA(I,J))
      C3=(1./(VA(I-1,J)*VA(I,J)))+(1./(HA(I,J-1)*HA(I,J)))
      NODE(I,J)=(C1+C2)/C3
   17 IF(ABS(TEST-NODE(I,J)) .GT. TOL)IT=1
   18 CONTINUE
C
C     **AT I=1, DELTA FI / DELTA Y = 0
C
      DO 19 J=2,M-1
   19 NODE(1,J)=NODE(2,J)
      IF(IT .EQ. 0)GO TO 21
   20 CONTINUE
   21 WRITE(6,22)L
   22 FORMAT(//////5X,'FINAL FLOW NET DISTRIBUTION AFTER',I3,' ITERATIO
     1NS'///)
      DO 23 I=1,N
   23 WRITE(6,24)(NODE(I,J),J=1,M)
   24 FORMAT(15F8.2//)
   25 STOP
      END
```

RELAXATION OF A FLOW NET FOR A GIVEN SET OF BOUNDARY CONDITIONS

NODES TOLERANCE=0.05

SIZE OF MESH ARMS

```
0.25  0.25  0.25  0.25  0.25  0.25  0.25  0.25  0.25  0.25  0.25
1.00  1.00  1.00  1.00  1.00  1.00  1.00  1.00  1.00  1.00  1.00
1.00  1.00  1.00  1.00  1.00  1.00  1.00  1.00  1.00  1.00  1.00
1.00  1.00  1.00  1.00  1.00  0.50  0.50  0.50  1.00  1.00  1.00
1.00  1.00  1.00  1.00  0.55  0.55  1.00  1.00  1.00  1.00  1.00
1.00  1.00  1.00  1.00  1.00  1.00  1.00  1.00  1.00  1.00  1.00
1.00  1.00  1.00  1.00  1.00  1.00  1.00  1.00  1.00  1.00  1.00
1.00  1.00  1.00  1.00  1.00  1.00  1.00  1.00  1.00  1.00  1.00
```

INITIAL NODE INFORMATION MATRIX

100.00	87.00	73.00	55.00	31.00	0.	0.	0.	0.	0.	38.00	61.00	77.00	89.00	100.00
100.00	87.00	73.00	55.00	31.00	0.	0.	0.	0.	0.	38.00	61.00	77.00	89.00	100.00
100.00	89.00	76.00	59.00	36.00	0.	0.	11.00	15.00	22.00	47.00	66.00	80.00	90.00	100.00
100.00	91.00	82.00	69.00	53.00	30.00	27.00	38.00	44.00	52.00	64.00	76.00	85.00	93.00	100.00
100.00	95.00	90.00	84.00	76.00	68.00	66.00	69.00	72.00	77.00	82.00	87.00	92.00	96.00	100.00
100.00	100.00	100.00	100.00	100.00	100.00	100.00	100.00	100.00	100.00	100.00	100.00	100.00	100.00	100.00

FINAL FLOW NET DISTRIBUTION AFTER 5 ITERATIONS

100.00	87.42	73.16	55.14	31.17	0.	0.	0.	0.	0.	38.12	61.31	77.19	89.36	100.00
100.00	87.42	73.16	55.14	31.17	0.	0.	0.	0.	0.	38.12	61.31	77.19	89.36	100.00
100.00	88.60	75.58	58.93	35.71	0.	0.	10.98	15.36	22.01	47.62	66.10	79.74	90.52	100.00
100.00	91.41	81.63	69.30	52.72	29.73	27.12	37.90	44.40	51.83	64.32	75.77	85.19	93.00	100.00
100.00	95.41	90.24	83.92	76.16	67.98	66.05	69.12	72.53	76.62	82.11	87.53	92.26	96.32	100.00
100.00	100.00	100.00	100.00	100.00	100.00	100.00	100.00	100.00	1000.0	100.00	100.00	100.00	100.00	100.00

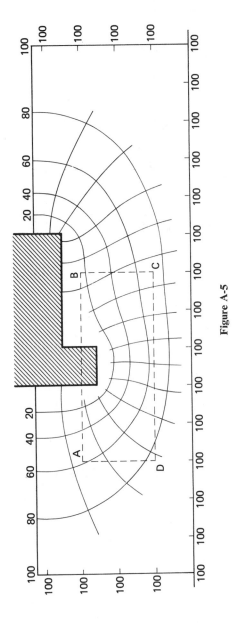

Figure A-5

SYMBOLS

A	area
b	channel bottom width
C	Chezy's coefficient
c	surface wave celerity (wave speed)
C_{HW}	Hazen-Williams coefficient
$°C$	degree Celsius
D	diameter, hydraulic depth
d	water depth
E	energy per unit weight of water (energy head)
e	roughness height
E_b	modulus of elasticity (bulk)
E_p	modulus of elasticity
E_s	specific energy
e	efficiency
EGL	energy gradient line
F	force
f	friction factor
F_s	specific force
g	gravitational acceleration

H	total head
h	elevation head
h_L	head loss
h_f	friction loss
HGL	hydraulic gradient line
I	moment of inertia
K	coefficient of permeability, coefficient of energy loss
L	length
l	liters
M	moment, momentum and mass
m	mass
M, N	exponents of the gradually varied flow function
N	newtons
n	Manning's coefficient
N_F	Froude number
N_R	Reynolds number
N_W	Weber number
$NPSH$	net positive suction head
P	wetted perimeter, pressure
P_i, P_o	input, output power
Q	flow rate (discharge)
q	discharge per unit width of channel
R	radius
r	variable radius
R_H	hydraulic radius
S_f	energy slope
S_c	critical slope
S_o	channel slope
S_w	water surface slope
sp. gr.	specific gravity
T	temperature
T	torque
t	time, thickness of pipe wall
T_r	return period
x, y	coordinate axes
u, v, w	velocity in x, y, z direction
V	mean velocity
Vol	volume
W	weight

σ	surface tension, cavitation parameter
α	porosity, energy coefficient
β	vane angle, momentum coefficient
ρ	density
μ	viscosity
ν	kinematic viscosity
γ	specific weight
τ	shear stress
ω	angular velocity

INDEX

Depression storage, 314
Depth:
  critical, 158, 160, 164, 170, 180
  hydraulic, 159
  initial, 163, 165
  normal, 149, 168, 170
  sequent, 163, 165
  sub critical, 163
  super critical, 163
  uniform, 149, 168
  water, 149, 170, 183, 186
Diagram:
  free body, 147
  polar vector, 121
  specific force, 167
  velocity vector, 120, 121
Diffusor, 67
Dimensional analysis, 200
Direct runoff, 314, 320
Discharge of model, 281
Displacement:
  longitudinal, 185
  radial, 185
Disturbance, 185, 186, 193
  in bend, 195
  wave pattern, 195
Disturbance center, 185
Drainage area, 318
Drainage system, 228
Drought period, 309, 322

E

Earth dam, 228
  drainage of, 228, 229
  embankment, 228
  homogeneous, 228
  phreatic line in, 228
  seepage through, 228
Earthquake force:
  on dam, 233, 234
  due to water mass, 233
Efficiency, 115
  of centrifugal pump, 123
  hydraulic, 152
  of jet pump, 128
  of multiple pumps, 137

Efficiency (*cont.*)
  of pumps in parallel, 137
  of pumps in series, 137
  overall, 115
Elasticity:
  bulk modulus of pipe, 104
  bulk modulus of water, 10, 104
  composit modulus, 106
  volume modulus of water, 10
Electrical resistivity, 217
Electric analog, 338
Electric current, 338
Electric potential, 217, 338
Electrode, 339
Electrolyte, 338
Elevation head, 48, 155
Energy:
  coefficient, 156
  consumption of, 231
  conversion, 231
  dissipation, 231, 253
  gradient line, 58, 63, 64, 109, 143, 161
  in hydraulic jump, 166, 167
  kinetic, 46, 55
  latent, 2
  loss of, 46, 143, 181
  in pipeline, major loss, 50
  potential, 46
  pressure, 46, 155
  slope, 61, 182, 183
  specific, 157, 182, 183
Energy dissipator, 257, 258, 259, 260
Energy equation, 49, 125, 131, 148
Energy head:
  datum, 49, 155
  elevation, 48
  potential, 48
  pressure, 48, 63, 104
  total, 160
Equilibrium:
  of floating body, 36
  of pipe flow, 51
  static, 21, 29, 30
  steady state, 204
Equilibrium test of aquifer, 206
Equipotential line, 224, 228, 336
Equivalent hydraulic system, 214
Evaporation, 2